Unmaking the Bomb

CRITICAL ENVIRONMENTS: NATURE, SCIENCE, AND POLITICS

Edited by Julie Guthman and Rebecca Lave

The Critical Environments series publishes books that explore the political forms of life and the ecologies that emerge from histories of capitalism, militarism, racism, colonialism, and more.

1. *Flame and Fortune in the American West: Urban Development, Environmental Change, and the Great Oakland Hills Fire*, by Gregory L. Simon
2. *Germ Wars: The Politics of Microbes and America's Landscape of Fear*, by Melanie Armstrong
3. *Coral Whisperers: Scientists on the Brink*, by Irus Braverman
4. *Life without Lead: Contamination, Crisis, and Hope in Uruguay*, by Daniel Renfrew
5. *Unsettled Waters: Rights, Law, and Identity in the American West*, by Eric P. Perramond
6. *Wilted: Pathogens, Chemicals, and the Fragile Future of the Strawberry Industry*, by Julie Guthman
7. *Destination Anthropocene: Science and Tourism in The Bahamas*, by Amelia Moore
8. *Economic Poisoning: Industrial Waste and the Chemicalization of American Agriculture*, by Adam M. Romero
9. *Weighing the Future: Race, Science, and Pregnancy Trials in the Postgenomic Era*, by Natali Valdez
10. *Continent in Dust: Experiments in a Chinese Weather System*, by Jerry C. Zee
11. *Worlds of Green and Gray: Mineral Extraction as Ecological Practice*, by Sebastián Ureta and Patricio Flores
12. *The Fluvial Imagination: On Lesotho's Water-Export Economy*, by Colin Hoag
13. *The Low-Carbon Contradiction: Energy Transition, Geopolitics, and the Infrastructural State in Cuba*, by Gustav Cederlöf
14. *Unmaking the Bomb: Environmental Cleanup and the Politics of Impossibility*, by Shannon Cram

Unmaking the Bomb

ENVIRONMENTAL CLEANUP AND
THE POLITICS OF IMPOSSIBILITY

Shannon Cram

UNIVERSITY OF CALIFORNIA PRESS

University of California Press
Oakland, California

© 2023 by Shannon Cram

Library of Congress Cataloging-in-Publication Data

Names: Cram, Shannon, 1980- author.
Title: Unmaking the bomb : environmental cleanup and the politics of
    impossibility / Shannon Cram.
Other titles: Critical environments (Oakland, Calif.) ; 14.
Description: Oakland, California : University of California Press, [2023] |
    Series: Critical environments: nature, science, and politics ; 14 | Includes
    bibliographical references and index.
Identifiers: LCCN 2023018071 (print) | LCCN 2023018072 (ebook) |
    ISBN 9780520395114 (cloth) | ISBN 9780520395121 (paperback) |
    ISBN 9780520395138 (epub)
Subjects: LCSH: Radioactive waste sites—Cleanup—Washington (State) |
    Hanford Site (Wash.).
Classification: LCC TD898.12.W2 C76 2023 (print) | LCC TD898.12.W2 (ebook) |
    DDC 363.72/890979751—dc23/eng/20230503
LC record available at https://lccn.loc.gov/2023018071
LC ebook record available at https://lccn.loc.gov/2023018072

32 31 30 29 28 27 26 25 24 23
10 9 8 7 6 5 4 3 2 1

Contents

	Introduction: On Telling Impossible Stories	1
1.	Tender	17
2.	Anatomy of a Phantom	41
3.	Rational Mutants	67
4.	Body Burden	87
5.	Trespassing	107
	Conclusion: Here, in the Plutonium	120
	Acknowledgments	131
	Notes	135
	References	181
	Index	207

Introduction

ON TELLING IMPOSSIBLE STORIES

Even now, I am tempted to tell you a different story. One about a dry, windswept place called Hanford in eastern Washington State where they used to make plutonium. I would describe the way it smells there (like warm sagebrush and dust) and paint you a picture of aging reactors glinting in the sun. I would tell you that, once upon a time, this was the beating heart of American nuclear weapons production and now it is one of the most contaminated places on Earth.[1] Now it houses the majority of the nation's high-level nuclear waste.[2] Now it is engaged in one of the largest environmental cleanups in human history.[3]

If I told you that story, I would do my best to impress you with the scale of the problem: 56 million gallons of radioactive sludge stored in leaky underground tanks; more than seventy square miles of toxic groundwater; hundreds of contaminated buildings, including nine reactors and five chemical processing plants that dumped about 450 billion gallons of liquid waste directly into the soil.[4] I would mention that even the tumbleweed is radioactive there. So are the coyotes and cliff swallows and rabbits.[5] I would tell you that it is actually someone's job to collect contaminated rabbit droppings and dispose of them in a burial ground on-site.[6]

And I would emphasize the waste's unruliness: how it ignores the No Trespassing signs that mark Hanford's official edges, how it escapes on wind and water currents, stows away in dust devils and groundwater plumes.[7] I would tell you that those plumes flow into the wide, blue Columbia River, which arches around the site for fifty miles, and I would add that this river was once the most radioactive in the world.[8]

I would say all of this in order to make the case for cleanup. That is really the point of the story I am tempted to tell: first to inspire concern about Hanford and its errant wastes, then to offer a solution.

There is a contaminated place.
It is hazardous to human health and the environment.
We need to clean it up.

It's not a hard sell. Even if you hadn't heard of Hanford before you opened this book, you already knew the plot. Maybe you thought about Fukushima as you read. Or the *Exxon Valdez*. Maybe you pictured hazmat suits and birds sticky with oil. I admit, the temptation to tell you a story that you already know has something to do with its familiarity. There is a sure-footed confidence to shared narrative. It can be easier to tell a recognizable tale.

Plus, I've had a lot of practice—I have been telling that story for almost twenty years. First, as a young canvasser going door to door for a political campaign about Hanford's waste and then as a graduate student studying the cultural politics of its cleanup. Later, as a university professor teaching classes about environmental health, and as a member of the Hanford Advisory Board writing policy advice about remediation.[9] That story is woven into the fibers of my personal and professional life. I am invested in it. I care about the people and places within it. I tell it in order to advocate for a more just future.

And yet. For most of the nearly two decades that I have been telling that story, I have struggled with a sense of duplicity. It's not that the story is untrue, per se, it's just not . . . enough. It is guilty of critical omissions and, as its narrator, I am complicit in those erasures.

First, and most fundamentally, for example, Hanford will never be cleaned up, if by clean we mean free of its waste. In fact, when remediation is officially complete many years from now, the majority of Hanford's waste will remain on-site.[10] There will still be plutonium in the soil and carcinogens in the groundwater. There may still be radioactive rabbits hopping around.

The word *remediation* derives from *remedy*, meaning to heal or to cure.[11] But the only absolute cure for Hanford's contamination is time. It will take hundreds of thousands of years for its radioactive light to go out.[12] You can't destroy plutonium or dilute it until it ceases to be dangerous. Even small particles of it, taking flight on the wind and landing in a lung, can cause cancer.[13] Cleanup does not attempt to eliminate such particles altogether, for that is not possible. Instead, it promises to monitor and contain these and other toxic materials for a "reasonable" amount of time (according to federal regulations, that period ranges from one to ten thousand years).[14]

Second, cleanup-as-containment does not necessarily mean building physical barriers around Hanford's waste or separating it from the environment. Although some remedies involve material interventions (i.e., removing contaminated soil from one area of the site and reburying it in another, more secure, area of the site), others employ "nonengineered instruments" known as "institutional controls."[15] No Trespassing signs are institutional controls. So are land use designations that determine what kinds of activities may occur on-site.[16] Institutional controls, in other words, often manage humans rather than waste.

Third, clean does not mean uncontaminated in U.S. environmental policy. Instead, places like Hanford are considered remediated when they can demonstrate an "acceptable risk" of exposure-related cancer among the imagined communities that may live, work, and/or spend time there in the future.[17] Clean, therefore, is not defined by the *absence* of waste, but by the *relationship* between waste and the body. Remediation measures and manages that relationship—titrating environmental contamination with probabilistic human activity in order to achieve legally compliant levels of disease.[18]

That temptingly simple version of the story I have been telling for years fails to address these conditions of possibility. Quite the opposite: it frames cleanup as a neutralizing force, implying that one day there will be an "after" to atomic violence. That story does not explain how (some) injury and death is built into cleanup's very logic—how, from a policy perspective, safe is synonymous with reasonable harm.

This basic problem—how to talk about Hanford—has long been a source of tension in my work. As an academic, I find it useful to break the story down and consider its parts, exploring its embedded assumptions.

There is a contaminated place.

What do we mean by *contaminated*? How should we understand this consequential category in a world saturated with industrial carcinogens and structural inequalities that amplify their effects? What situated histories have produced both the materials themselves and their entangled forms of reckoning?

It is hazardous to human health and the environment.

What do we mean by *hazardous* and, for that matter, *health*, on an unevenly contaminated planet? What specific bodies, environments, and rationalities have come to define those terms, and what is it like to live their definitions?

We need to clean it up.

What do we mean by *cleanup*, and how is that meaning informed by our answers to the preceding questions? In other words, how does remediation reflect the powerful social relations of contamination, health, and hazard?

Unmaking the Bomb investigates these politics of impact and remedy in the atomic age. It explores how frames like exposure and protection, risk and cancer, reason and practicability recognize (and fail to recognize) contaminated life. And in considering how such forms of recognition have come to be, it asks how they could be otherwise.

So too, this book represents an ongoing effort to understand my own relationship with risk and remediation. In writing this introduction, for example, I went in search of the old research notebooks that I had saved from my master's thesis. The younger self I found in those pages posed eager and impatient questions, determined to identify *the* answer to Hanford's nuclear waste problem. One set of notes from 2005 described a moment in a public meeting when I strode up to the microphone and demanded that Washington's Department of Ecology director remediate Hanford *immediately* (my notes include the phrase, "I gave him hell.").

Reading this description today makes me blush, but I am touched by it as well. One of the gifts that comes with long-term research in a particular place is that field notes and free-writes, old calendars and interview transcripts, represent more than a record of research practice and developing ideas. They also illustrate the complex processes and relations of becoming with that place through time. *Unmaking the Bomb* is informed by my efforts to grapple with Hanford's power-laden logics, my struggle to negotiate the boundaries between research and activism, and my need to reckon with the conditions of living and dying in the nuclear era. In many ways, therefore, this book tells a personal story that weaves Hanford's histories together with my own.

I have grown up with Hanford in ways that I could not have imagined when I started this project. Since that time, every member of my immediate family has gotten cancer, and both of my parents have died of it. My father was diagnosed at age fifty-five (and died the same year), soon after I began canvassing door to door about Hanford in 2004. My mother was diagnosed at sixty, while I was completing dissertation fieldwork about cleanup in 2012. I was diagnosed at thirty-three in 2014, just weeks after my mother passed away. I spent my final year of graduate school writing my dissertation and applying for professorships, in addition to completing five months of chemotherapy and two major surgeries that removed my breasts, ovaries, fallopian tubes, and a handful of lymph nodes. My sister was diagnosed at age thirty-one, a few weeks before I finished my PhD. We went to her first oncology appointment together, the morning after I graduated. She was diagnosed again four years later and successfully completed treatment for a second time, as I was finishing the first draft of this book.

I include this personal context because my family's experience with cancer has informed how I think about Hanford, the nuclear industrial complex, and the daily politics of toxicity. It matters that the research and writing for this book took place between doctor's appointments, surgeries, chemotherapy infusions, and funerals for two of the people that I loved most. It matters that I read studies about nuclear fallout while waiting for biopsy results and that I traced the history of U.S. toxics policy while leaden with grief. It matters that my research files contain declassified maps of contamination in the communities surrounding Hanford, and that those maps include my mother's hometown.[19] It matters that I want to know what caused my family's cancer. And it matters that I will never be able to fully answer that question.

For although most environmental quality standards are based on statistical models of carcinogenic hazard, it is nearly impossible to identify when an individual instance of cancer results from daily life in a contaminated environment. Cancer remains the primary risk factor driving environmental legislation in the United States—it is used to establish baselines for acceptable toxicity concentrations in air, water, soil, vegetation, and bodies and to determine if those contaminants have exceeded permissible limits. More than any other, this disease has informed the categories we use to define and regulate environmental health, from air pollution in Los Angeles to nuclear waste at Hanford.[20] However, when cancer shapeshifts from a risk metric into a living and dying body, its origin story becomes largely unrecognizable.

Instead, individuals with cancer are left to wonder how they could possibly have gotten it. Causation is often framed as a personal failure: the unfortunate and even embarrassing result of poor diet, not enough exercise, too much stress, and so on. I remember feeling this acutely one day when a friend who had learned about my family's history said to me, "Jeez, what have you guys been doing wrong?"

As anthropologist Lochlann Jain argues, even beribboned campaigns that raise awareness and celebrate survivors often narrate cancer through individual struggle and personal accomplishment instead of potential links with environmental exposure. "Cancer becomes a passively occurring hurdle to be surmounted by resolve," they write, "rather than the direct result of a violent environment."[21] Ironically, efforts to mitigate such violence through regulation and remediation often reiterate this disconnect even as they seek to resolve it. This and other paradoxes integral to environmental cleanup are at the heart of this book.

As difficult as it is to admit to my younger self, I do not attempt to solve Hanford's nuclear waste problems here (at least, not in the totalizing way I once imagined). In fact, much of my research explores the structural impossibilities of doing that very thing. Instead, I position ambiguity and contradiction as avenues for critical discussion, rather than as roadblocks to it. I suggest that uncertainty is more than an absence of knowledge, and I attend to the social relations of not knowing.[22] Finally, I make the case that improving the terms of cleanup means taking impossibility seriously—asking seemingly basic questions like these: How can we regulate a waste form that will long outlast the United States and its regulatory structures? Whom does reasonable exposure protect, and whom does it harm? What does it mean to safeguard individual bodies with regulations that only envision disembodied statistical aggregates? And how have politically and economically tenable solutions come to define the problems of nuclear cleanup and safety?[23]

When writing this book I had the opportunity to interview the former Department of Ecology director whom I "gave hell" in 2005. We had a nice conversation. He was generous and helpful, offering suggestions when I complained about the narrative challenges that Hanford presents. "Here's what I think you should write about," he told me. "This [nuclear cleanup] is not only a test of the United States, this is a test of our species. The genie is out of the bottle, and there's no putting the genie back in. Well," he paused and pointed to an old picture of Hanford on the table between us, "this is the legacy of that genie. This is a test of our society. Are we really willing to do what it takes to remedy this situation?"[24]

His question has stayed with me. In fact, since then I have noticed it being asked frequently, albeit in different forms. It emerges when Hanford managers describe political gridlock and budgetary constraints, and when community organizers advocate for better waste treatment protocols. Indeed, in many ways, "Are we willing to do what it takes?" is a very practical question. It invites iterations: How much money is required to make this project work? What kind of regulations would be necessary? Do "we" as individuals, nations, and communities have the resources to make cleanup happen?

However, such questions also imply that the bomb can, in fact, be unmade—that if only there were bigger budgets, better technologies, and greater public interest, this situation could be remedied. Yet these same people also recognize the regulatory impossibilities of nuclear waste. They acknowledge that the genie has already left the bottle and there is no putting it back.[25]

To be clear, when I say impossible, I mean both the material challenges of Hanford's cleanup as well as the normative stories we tell about it. I mean that multimillennial waste will inevitably exceed its physical and institutional containers, and that administering eternity has unthinkable, science-fiction-like qualities.[26] But I also mean the powerful conditions and contexts that define unthinkability itself. I mean the social politics that designate some impacts as reasonable and others as inconceivable, allowing cleanup to distribute survival unevenly.[27] By impossible, therefore, I mean both the concrete and constructed realities of contaminated life and the oft-blurred boundaries between the two.

Also, when I say we need to take impossibility seriously, I am not making a case for inaction. On the contrary, I see equitable, long-term waste management as essential to a socially and environmentally just future at Hanford. Instead, I argue that improving the terms of cleanup means asking better questions. Rather than "Are we willing to do what it takes?," we should be asking: What are the politics of our actions? What are the conditions in which remediation is designed, embodied, enacted, and understood? What infrastructures give these actions power, and what does this tell us about our capacity to create positive change? For that matter, what would positive change look like? Positive for whom? Unmaking the bomb requires much broader forms of critical engagement. It insists that we reckon with the very meaning of nuclear impact while acknowledging that its unmaking will never be complete.[28]

One of the distinct challenges I have found in narrating Hanford is learning to think with and against its epistemic frames.[29] Toxicity, for example, is the product of economic, technoscientific, and regulatory practices that have made some environmental exposures perceptible and many others imperceptible in the name of industrial development.[30] As historian Evan Hepler-Smith argues, even the molecular structures that inform U.S. toxics policy were originally designed to facilitate industrial chemical production and its bureaucracies.[31]

At the same time, toxicity is also the product of community organizing and knowledge making that reimagines such logics in the service of social and environmental justice. Some of these efforts leverage the same molecular bureaucracies to resist their structural invisibilities, documenting toxic residues that would otherwise remain unmeasured and unseen.³² Others look beyond the molecular to consider how toxicity "functions as a proxy for a range of cultural, economic, or infrastructural instabilities that are, indeed, something 'toxic' but are far more complicated and difficult to identify."³³

Perhaps most importantly, toxicity is produced in and through the impossibility of its absence. There is no outside to industrial production and its inequitable body burdens.³⁴ Nuclear waste is thus inseparable from social formations of race, indigeneity, gender, class, disability, and others. Indeed, as historian M. Murphy writes, the body itself represents "a collective binding of profoundly uneven relations of porosity to exposure: my vulnerability to injury is entangled with your comfort. The side effect accompanies the treatment. I am kept alive even as I am being killed."³⁵ *Unmaking the Bomb* thinks with and against cleanup as a distinct form of toxic embodiment. And in engaging the impossible stories of nuclear waste, it offers an incomplete remedy.

When I introduce Hanford as a topic in my classes, I often begin by asking students to close their eyes and picture the bomb. They sit for a few moments, noting the first image that comes to mind, and then draw it on a piece of paper. Next, I invite them to hold up their drawings, look around the room, and consider their classmates' nuclear imaginations. Invariably, as they twist their heads and scan one another's sketches, the space fills with soft murmuring and scattered laughter. It is the sound of dawning recognition: almost all of them have drawn the same thing.

It's no accident, of course, that my students see mushroom clouds when I ask them to picture the bomb. Atomic weapons promise protection by threatening catastrophe, envisioning a devastation so great as to make war unthinkable. As historian Paul Edwards argues, in Cold War geopolitics, simulated disaster "became more real than the reality itself, as nuclear standoff evolved into an entirely abstract war of position. Simulations— computer models, war games, statistical analyses, discourses of nuclear strategy—had, in an important sense, more political significance than the weapons that could not be used."³⁶

Imagining the mushroom cloud, therefore, was essential to the bomb's very utility. Its roiling, top-heavy form represented the ever-present potential for destruction, a haunting vision of what could be. Americans learned to live in its spectacular/speculative shadow, practicing duck-and-cover drills in school, building fallout shelters in basements, and watching cinematic renderings of atomic tests on TV. The U.S. Office of Civil Defense (OCD) argued that such visuals were critical to emotionally preparing the nation for nuclear crisis and unifying support for Cold War policies. The challenge, OCD officials explained, was achieving the right balance between terror and fear, framing mental discipline as necessary to surviving the atomic age.[37]

So too, nuclear pedagogy required that Americans accept several distinct contradictions. First, that avoiding atomic war meant actively preparing to achieve it: developing massive industrial economies, building production facilities in almost every state, and contaminating vast areas of land. Second, that in order to prevent foreign attack on American soil, the federal government would need to perform its domestic equivalent. In fact, by the end of the Cold War, the United States had atomic-bombed itself more than nine hundred times in the name of its own protection.[38] Third, that nuclear safety could coexist with nuclear risk: that radiation protection could be defined by reasonable exposure.

It is notable that my students—most of whom were born more than a decade after the Cold War ended—have retained its curated imagery. Though I know it is coming, that moment in class when everyone holds up their mushroom clouds still takes my breath away. We spend the next hour discussing the bomb's other spatial and temporal forms. We talk about waste and slow violence, we examine maps of nuclear fallout, we consider the embodied impacts of production and testing.[39] Despite this added context, however, students tell me that the mushroom cloud remains central, even definitional, to their mental picture of the bomb. And I get it: after all of these years unsettling the image, it's usually the first one that pops into my head, too.

As literary scholar Jessica Hurley points out, the mushroom cloud often remains a stubborn abstraction in nuclear criticism as well. Academic engagement with the bomb has been "held somewhat captive" by the existential threat of world-ending war, she writes, an epistemological condition known as the "nuclear sublime."[40] Yet "when we speak of the nuclear as an always-absent referent, as that which we cannot think," she continues, we risk "being blinded to the negotiations of power, wealth, status, and vulnerability that are constantly in play around nuclear and contestably nuclear things, from bodies and rocks to highways and international treaties."[41] Instead, Hurley proposes the "nuclear mundane" as a more useful heuristic for thinking with the atomic age. By "focusing on the environmental, infrastructural, bodily, and social impacts of nuclear technologies and the politics that prioritize them," she argues, the nuclear mundane reimagines the bomb "as continuous with a set of militarized infrastructures rather than as their exceptional end point."[42]

In some ways, cleanup represents an effort to understand atomic impact beyond the mushroom cloud. As Secretary of Energy Hazel O'Leary wrote in 1997, "the full story of the splitting of the atom has yet to be written" and one of its "biggest missing pieces" is the "environmental legacy of nuclear weapons production in the United States."[43] Reckoning with that legacy, she argued, would require more than dismantling the bomb's physical infrastructures; it would also mean crafting a broader narrative about the nation's nuclear history.

This, she acknowledged, would be no small task. By the end of the Cold War, U.S. atomic weapons production had become a trillion-dollar industry.[44] As Manhattan Project physicist Niels Bohr predicted in 1939, making the bomb had "turned the United States into one huge factory."[45] Indeed, by the time the Soviet Union formally dissolved in 1991, the U.S. nuclear complex officially occupied thousands of square miles and maintained massive weapons-based economies surrounding production facilities in Oak Ridge, Tennessee; Richland, Washington; Los Alamos, New Mexico; Aiken, South Carolina; Amarillo, Texas; Idaho Falls, Idaho; Jefferson County, Colorado; and many other smaller sites.[46]

This nationwide factory was never designed to stop making nuclear weapons. As former Department of Energy (DOE) adviser Bob Alvarez explains, peace was a "profoundly disruptive thing to a system that had never envisioned stopping. The United States and Russia, and especially the people running their nuclear weapons industries, never thought about stopping and what that means. There were no contingencies. They just thought this would go on forever."[47]

Thus, as the Cold War came to a close, the federal government began the previously unthinkable task of postproduction accounting. In 1989 Secretary of Energy James Watkins created the Office of Environmental

Restoration and Waste Management with the express purpose of "mitigating the risks and hazards posed by the legacy of nuclear weapons production."[48] In 1993 O'Leary (Watkins's successor) launched a department-wide Openness Initiative, renaming the Office of Classification as the Office of Declassification and revealing thousands of wartime human radiation experiments and intentional contaminant releases.[49] "The Cold War is over," she said at a December press conference that year, "we're coming clean."[50]

In 1995 the National Defense Authorization Act required that DOE provide a detailed analysis of weapons-based waste and contamination. This inspired a series of reports with titles like *Closing the Circle on the Splitting of the Atom: The Environmental Legacy of Nuclear Weapons Production and What the Department of Energy Is Doing about It, Estimating the Cold War Mortgage: The 1995 Baseline Environmental Management Report,* and *Linking Legacies: Connecting the Cold War Nuclear Weapons Production Processes to their Environmental Consequences.*[51] In each, DOE assessed the material remains of a national security strategy that had privileged the immediate threat of nuclear attack over the slow violence of environmental contamination. The result was a distinct "geography of sacrifice," requiring a nationwide cleanup effort likely to cost more than the arsenal itself.[52]

However, even as they emphasized the magnitude of environmental impact, these reports also positioned cleanup as the manageable second half of the nuclear production cycle. In *Closing the Circle,* for example, Assistant Secretary of Environmental Management Thomas Grumbly likened the nation's atomic story to the "full sweep of a clock face."[53] At noon, he wrote, the world's first nuclear bombs were born, followed by Cold War geopolitics and weapons manufacture from 2:00 p.m. until 5:00 p.m. Early evening marked the war's end along with the majority of its weapons programs: the Berlin Wall crumbling at around 5:30 p.m., in time for U.S. nuclear remediation to begin at 5:45 p.m. Looking forward, Grumbly envisioned the final six hours and fifteen minutes of the nuclear project in which contaminated landscapes would be scrubbed clean and radioactive wastes stored securely. At midnight, he imagined, the hands of nuclear time would return to their original position, sealing the endeavor with a neat click.[54]

Selecting a circle as the shape of atomic progress pointed to several DOE mandates for remediation. First and most evident, for example, was that of containment. By "closing the circle," the agency was rhetorically re-sealing Pandora's box, asserting control over runaway radionuclides and radically uncertain futures. So too, circular imagery portrayed nuclear production as cyclical, suggesting that remediation activities would restore contaminated land to its original state. Finally, by envisioning the atomic age as a loop—expanding and retracting, moving forward while forever returning to itself—DOE framed remediation as not only possible but inevitable, as certain as the passage of time.

Such narratives belied the bomb's profound spatial and temporal scale. Though the end of the Cold War marked a significant shift from production to cleanup, it failed to address the broader rationalities of reasonable harm. Instead, DOE's remediation efforts offered another curated vision of nuclear life, highlighting particular hazards while rendering others invisible.

During that first discussion about Hanford in my classes, most students react with surprise. This response often has less to do with the scale of contamination and more to do with the fact that they have never heard of the site. Like me, many of my students are from Washington and/or have spent much of their lives in this state. How, they ask, is it possible that they did not know it houses most of the nation's high-level nuclear waste? Or as one person put it, "How is it that I know more about the Kardashians than this place?"

I like this question. It makes for a generative discussion about the structural invisibilities of environmental contamination and the social politics of storytelling. It also creates a shared experience around a largely unfamiliar topic—making that lack of familiarity an interesting and important story in and of itself. One year a group of students even designed their final project around the question, producing a podcast episode called, "Why Don't People Know about Hanford?"

Yet that initial surprise is also woven through with a set of assumptions about what knowing a place like Hanford might mean. As with mushroom clouds, atomic landscapes come with culturally produced expectations about hypervisible transformation—what historian Gabrielle Hecht describes as "nuclear exceptionalism."⁵⁵ Students often assume that the site's contamination must be obvious, its effects immediate and legible. They make jokes about Blinky, the three-eyed fish from *The Simpsons*.⁵⁶ I remember my own introduction to Hanford as a young twenty-something too and the disorientation I felt during my first tour of the site. I couldn't believe how *normal* everything looked.⁵⁷

Of course, as Hecht points out, nuclear exceptionalism is also "full of contradictions."⁵⁸ The same civil defense materials that emphasized the bomb's catastrophic potential rendered it banal in the same breath. "Government propagandists assured citizens that simple gestures offered protection if the bombs did fall," she writes. "American schoolchildren could take refuge under their desks, sang Bert the Turtle in the famous 'Duck and Cover' ditty. Fallout shelters promised the perpetuation of suburban lifestyles in the event of nuclear war."⁵⁹ Civil defense, in other words, "did not resolve the problem of the bomb," anthropologist Joseph Masco argues, but instead made its contradictions integral to nuclear safety. It asked Americans to accept a narrative of protection "founded simultaneously on total threat and absolute normality—with the stakes being nothing less than survival itself."⁶⁰

Such logics extended to regulatory design in Cold War nuclear production. One of the challenges in developing exposure standards for a growing industry was that there was no entirely safe amount of radiation. Especially when it came to genetic mutation, most scientists at the time agreed that even the smallest dose could "injure the hereditary materials."⁶¹ So too, low-level exposures produced disparate effects that could take decades or generations to manifest, making it difficult to establish causality. How, then, to design safety standards without a specific threshold for safe exposure?

After some debate, the U.S. National Committee on Radiation Protection (NCRP) adopted a conceptual framework called *permissible dose* in 1946.⁶² Rather than claiming no impact, it placed the potential costs of radiogenic harm alongside the potential benefits of nuclear technology and identified a "reasonable" balance between the two. In particular, permissible dose reimagined safety through the probabilistic risk of injury that an "average individual" would be willing to accept in the name of atomic progress.⁶³

Nuclear production therefore initiated a series of profound shifts in the American policy landscape. Most importantly, as environmental historian

Linda Nash argues, it introduced the concept of risk into U.S. public health regulation.[64] With roots in defense planning and economics, risk functioned as both an analytical technique and a political philosophy. It understood well-being as a unit of market-based growth, weighing the benefits of development against the costs of disease.

Whereas previous policies had "assumed that anthropogenic pollution should not be allowed to materially affect health," Nash writes, risk framed exposure as "something to be managed, not avoided."[65] Concerns about fallout, for example, needn't spell an end to weapons production and testing. "The question was not whether to pursue an activity, but how to minimize the chance of failure and the extent of collateral health effects."[66]

This administrative turn inscribed federal environmental protections with Cold War rationalities.[67] Rather than prevention, risk methodologies sought optimization: statistically informed trade-offs between genetic mutation and national security, cancer and profit, public health and GDP. Safety standards based in acceptable risk formalized the notion that contamination was a necessary consequence of economic growth. They made reasonable harm a regulatory objective.

As such, Joseph Masco argues, "the nuclear arms race fundamentally altered—indeed placed in opposition—the concepts of 'national security' and 'public health.'"[68] Injury was not antithetical to the nation's prosperity in this formulation; it was essential to it—a routine expectation of technocratic modernity. No longer simply "an absence of disease," he writes, the atomic age recast health as a "graded spectrum of dangerous effects now embedded in everyday life." It defined survival as a form of "incipient death," making living a question of degree.[69]

By the time Hanford's cleanup officially began in 1989, acceptable risk was integral to U.S. remediation policy. It determined which sites were placed on Superfund's National Priority List and provided the "basis for action" in response to contaminant releases.[70] It informed the scale and approach of required interventions and offered a metric for determining when cleanup was complete. Remediating the nuclear complex, therefore, effectively reproduced some of the bomb's foundational logics. Just as the mushroom cloud envisioned protection through the probabilistic specter of disaster, cleanup offered a brand of safety haunted by reasonable toxicity.

I have perhaps never been so aware of Hanford's narrative challenges as when writing this book. How to capture a place, a problem, a relationship in all of its multilayered complexity? How to make the case for a more just and equitable cleanup that both highlights and unsettles its impossibilities? The project has felt at once too big and too small; the book, its own leaky container.

In the end, I selected but a handful of examples for thinking with and against the story of cleanup. Most center the body as a lived practice and regulatory device tasked with mediating environmental impact.[71] Feminist scholars have long argued that the body does not "end at the skin" but is instead constituted of and by histories, standards, and structures of power.[72] This is more than a metaphor: residing on Earth literally means becoming an industrial product, inhaling and absorbing, circulating and secreting permissible dose.[73] Yet though the body is always already altered by such exposures, it is not wholly defined by them. To be alive is to comprise and exceed life's social forms, to be both coded and uncontained.[74] *Unmaking the Bomb* inhabits these contradictions.

So too, this book pays close attention to the politics of narrative itself. For if cleanup speaks in the language of reasonable harm, then what can it possibly say of recovery?[75] I wrote this book because I wanted to investigate the social origins of Hanford's vocabulary, to trace how risk and remedy became inextricably entwined. I wrote it because I was tired of DOE's definition of clean performing a kind of rhetorical closure, serving as endpoint rather than opening. I wrote it because I longed to unwind such categories and consider their threads in search of alternative ways of telling.

The chapters that follow engage in a kind of narrative contamination, toggling back and forth between cleanup's administrative frames and the stories that overspill them.[76] I investigate how the body-at-risk became a waste management tool, its damage produced and prevented in probability. I spend time with the "human receptors" that occupy Hanford's exposure scenarios and the anthropomorphic phantoms that inform its dosimetry. I consider the statistical calculus of causality in workers' compensation and the nonstatistical people who live with its effects. And in the process, I explore the uneven social relations that make toxicity a normative condition.

1 Tender

When I entered the Robotics Pavilion at the 2018 Waste Management Symposia (WMS), the first being I encountered was Valkyrie. She was tall and broad-shouldered, meatless yet muscular, the most humanoid of her mechanical companions. Staring up at her six-foot-two, three-hundred-pound frame, I was reminded of Iron Man in a white suit, complete with a glowing orb between her hard plastic breasts.[1] Unlike Stan Lee's comic book character, however, there was no face behind Valkyrie's helmet—just a whirring LIDAR sensor giving her the lay of the land. In fact, she was all sensors: nearly one hundred of them in her palms and fingers alone, along with "hazard cameras" in her torso and a collection of surveillance technologies in her head. One might even say that sensing was her superpower.[2]

Known as a "caretaker robot," Valkyrie belonged to a class of nonhuman assistants designed to work in "damaged or degraded" environments.[3] Her makers at NASA hoped that one day she would participate in disaster response, nuclear cleanup, weapons disposal, and the American effort to colonize Mars. Though different in shape and technique, Valkyrie's robotic kin in the pavilion shared her affinity for the extreme. They were bound by a common purpose to inhabit hazardous space and, ultimately, to thrive where humans could not.

For the first time in forty-four years, nuclear robotics was the featured theme at WMS, reflecting dual industry imperatives to penetrate burgeoning tech markets and address aging waste infrastructures. In the months leading up to the conference, promotional materials promised new approaches to old challenges. "Don't miss the opportunity to exchange ideas, technical information and solutions with 2,000 nuclear waste industry delegates from more than 35 countries," organizers insisted. "If you are an influencer in the industry, then WM Symposia is where you need to be."[4]

I had heard mixed reports about WMS from Hanford workers and government regulators over the years. Some described it as big and boring—not worth the time and expense to attend—while others gushed that it was a "who's who" of nuclear cleanup, not to be missed. Eventually, my curiosity won out and I paid the steep $1,500 registration fee. For good measure, I also signed up for an evening nuclear networking event and a day-long radiation risk assessment training for Superfund remediators. More than anything else, however, I wanted to spend time on the show floor—what WMS billed as an "opportunity-rich environment" with corporate vendors, live product demonstrations, and a virtual reality golf course for when you needed to "take a break from your busy schedule."[5]

As the date neared, conference organizers began posting anticipatory photos on Facebook with the caption, "Show Floor—behind the scenes edition." In the first, the Phoenix Convention Center's seven-acre basement was bare, its concrete floors gleaming under fluorescent lights. In the next, the same vast space was clothed in brand-new gray carpet, a bright orange and purple registration area awaiting attendees. In still another, a gigantic robotic arm reached out from the back of a semitruck beside stacks of cardboard boxes and a dark green ladder. "Who's excited??" the organizers posted. "We know we are! #WMS #Nuclear." As I scrolled through the images on my way to Arizona, I had to admit that I was, too.

I wasn't expecting to find a magical techno-fix at WMS, but I was looking forward to a few heady days on the cutting edge. At Hanford, things were old. Dangerously so. Its waste facilities were literally buckling under the weight of time. Just ten months earlier, the site had made international news when part of a storage tunnel collapsed, leaving a gaping radioactive hole in the earth and thousands of workers sheltering in place. That hole felt like a metaphor for the cleanup itself: one more instance of "improbable" nuclear escape, one more narrative of control caving in.

It wasn't just aging waste facilities that needed attention at Hanford; it was also basic essentials like roads, plumbing, and ventilation systems. There were broken water mains to mend, rotting electrical poles to replace, cybersecurity systems to upgrade. More and more, maintenance costs were competing with cleanup efforts. More and more, Department of Energy officials were reminding Hanford Advisory Board (HAB) members like me that resources were limited and our policy recommendations should reflect that reality.

Soon after the tunnel collapse, DOE held a meeting at the Richland Public Library explaining that budgetary constraints were forcing the agency to make difficult decisions. It cost about $700 million per year to ensure "minimum safe operations" at Hanford, for example, but Congress had only allocated $800 million to the Richland Operations office in 2018.[6] This meant that most of the money would have to go to "min-safe" requirements, leaving DOE far short of what it needed to meet remediation milestones.

Given reduced budgets moving forward, agency officials wanted to know how we (members of the public) would prioritize cleanup activities. Referencing a series of posters placed around the room, DOE staff described projects that needed completing along with their anticipated costs: $700 million for canyon remediation, $340 million to address the K reactors, $1.1 billion for cleaning up the Central Plateau, and, and, and. . . .[7] How, they asked, would we rank these monumental tasks on a scale from one (most urgent) to ten (least urgent)? What was more important: infrastructural upgrades or cocooning reactors? Leaking high-level waste tanks or deteriorating pools filled with cesium and strontium capsules?

Black Sharpies in hand, audience members moved from poster to poster, considering each project and assigning it a number. All around me, people murmured to one another about the scale, the expense, and the challenge of having to decide. I heard two older women exclaim over an image of rusty waste storage barrels, agreeing that surely *this* deserved a higher ranking. I listened to a middle-aged couple weigh the price tag for pumping and treating contaminated groundwater, debating value in dialectic relation with time.

It was exhausting, infuriating even, being asked to construct a fiscally conservative hierarchy of need. The exercise felt like a misdirection from the broader structural issue that nuclear waste management *could* be

Cleanup projects and their anticipated costs. Hanford site cleanup priorities public meeting, Richland, WA, June 7, 2017. Photo by author.

underfunded in the first place. DOE had a legal obligation to meet its cleanup milestones, which to my mind meant that Congress *must* provide compliance-level funding. Every single project was urgent. Every single project was a priority. I wrote the number one on each poster in the room.

In contrast with Hanford's crumbling infrastructures and limited resources, WMS promised novelty and excess. As the self-proclaimed "largest and most prestigious conference on radioactive waste management and disposal," it drew leaders and practitioners from around the world.[8] "We're at the cusp of a new day in technology and its impact on the nuclear waste industry," one corporate sponsor said of the event. "Some of the capabilities that were once seen as impossible are now a reality."[9] Or, as the glossy flyer I found tucked into my conference program declared: "Hello, future.... Everything is possible."[10]

And indeed, the notion that anything could be seemed the very point of WMS 2018. As I moved past Valkyrie through the Robotics Pavilion and then beyond to the broader show floor, I marveled at the newest offerings on display. There were drones hovering like mosquitoes beside informational booths and small boxy robots rolling across the carpet on tiny tank tread. There was a mechanical stick-insect building a tower of bright green blocks and a two-legged humanoid named Wanderer posing with a soccer ball. I watched a man with exoskeletal arms lift boxes with cyborgian strength and another with virtual reality goggles reach for something that only he could see. I started when Robomantis, a human-sized spider swaying back and forth on black and orange legs, suddenly raised one of them in my direction.[11]

There were waste storage and transport options of all shapes and sizes: clean white MacroBags that could hold more than twenty thousand pounds at a time, cobalt blue ATOM containers large enough to walk inside, lime green "smart drums" that could monitor their own contents and alert operators when they were about to leak. I watched a ScanSort machine use nuclear isotope spectroscopy to organize radioactive debris by disposal category and learned how a robotic sprayer system called Fraccess removed sludge from a wastewater tank (a hazardous component of fracking often performed by human scrubbers). Even the vendors' names—Perma-Fix, Radwaste Solutions, and Vigor—communicated a hopeful techno-fetishistic message.

The show floor felt like a mix of Wall Street, Silicon Valley, high school Robotics Club, and mechanical engineering convention. Everywhere, men in suits were talking, laughing, shaking hands, demonstrating product features, operating joysticks, and staring at their phones. Everywhere, corporate taglines were making bold promises:

AECOM: *Imagine it. Delivered.*

Veolia: *Resourcing the World.*

SECUR: *Orchestrating Calm Even When
You're Surrounded by Complexity.*

The excitement in that windowless convention center basement was overwhelming. Unfortunately, so was the smell.

The seven acres of brand-new carpet I'd seen in the WMS Facebook feed had infused the room with a sickly sweet chemical tang. It was the aromatic equivalent of sucking on a penny: an odor that coated the throat and lingered sharply on the tongue. There was no ignoring that smell, no growing accustomed to it over time. Within minutes, my head was pounding.

I had planned to spend four days on the show floor exploring the political economy of nuclear waste through a fine-grained ethnography of the space. I had spent my entire travel budget for the year and canceled my classes for a week just so I could be in that room. Instead, I could only muster twenty minutes before the pain in my head forced me to leave the show floor. Twenty minutes marveling at robots while simultaneously scanning the faces around me for signs that they also were uncomfortable. Twenty minutes racing about taking photographs that I hoped would jog my memory when it came time to write this chapter. Twenty minutes angry with myself for being so sensitive. Twenty minutes without having a single conversation.

Afterward, I returned to my Airbnb and closed the shades, but my headache pulsed on for the rest of the afternoon. As I lay there in the dark, kicking myself for what I was missing at the conference, I thought about another time when new carpet had driven me from a room.[12]

That was December 2013, at the Hanford House hotel in downtown Richland, where I'd come for a three-day HAB meeting. Like its namesake, the hotel was old, a faded symphony of brown and burnt orange with a retro ballroom christened for one of the region's atomic boosters and a space-aged antenna sprouting from the roof. The new carpet was part of a much-needed remodel.

After three hours at Hanford House that winter, I was sick: dizzy, nauseous, head pounding like a drum. When I raised the issue during our lunch break at a nearby cafe, other HAB members acknowledged the smell (some complained of headaches too), but said they planned to soldier on. I felt weak in the shadow of their endurance. If they could push through, why couldn't I? Maybe *I* was the problem that needed solving.

In the end, I was the only person who left the meeting. I packed up quietly, cheeks burning with embarrassment, and drove four blocks to the public library, where I called in via Skype. Walking out as everyone else settled back into their chairs for the afternoon felt irrational and paranoid. If this was my reaction to carpet, how could I be trusted with nuclear waste? Leaving seemed to prove that my hazard barometer was off.

At the same time, I was also distinctly aware of the irony lurking behind that fog of shame. I was leaving a meeting about the social politics of toxicity because the meeting space itself was toxic. I was calling in from four blocks away to strategize about ensuring a livable future at Hanford without addressing the unlivable present in that room. Rather than using this as an opportunity to talk about the daily life of exposure (ostensibly the central problematic of cleanup), I had sidestepped the elephant in the room. I had apologized for my unreasonable body. I had internalized the problem of harm.

Even worse, this was the very kind of conversation that I wanted the board to have. I had spent the past four and a half years getting to know Hanford workers who, after being exposed on the job, were struggling to navigate the ambiguities of environmental illness. Their bodies had been changed by the work itself, they said in frustration; normal sensitivity metrics no longer applied. One worker described how a lit cigarette or a neighbor's perfume could initiate respiratory failure, sending her to the emergency room. Ordinary tasks like grocery shopping now required that she envision the invisible, anticipating geographies of potential hazard wherever she went. And one of the most confounding things about her exposure, she told me, was that its effects were inconsistent: some days, she was more sensitive than others. This capriciousness weakened her fight for workers' compensation—when she felt better, she undermined her own cause. Unremarkable (but smelly) activities like pumping gas had become evidentiary moments, the simple act of living an exercise in credibility. Recovery was a contradictory act. "It makes you look like a liar," she said.[13]

I had heard many such stories from workers over the years: anecdotes about headaches and sore throats, difficulty breathing and memory loss—some told in laughter, others in tears. There were self-described "quirks" like needing to run from the room when a family member sprayed down the kitchen counters, or a left eye that swelled shut in response to certain smells. There were tales of frustration and regret, dark humor and self-doubt. There was anger and embarrassment at not being believed.

While it was not my intention to establish equivalency between new carpet and nuclear waste, I did want the board to talk about the sensory politics of exposure. I wanted us to explore the powerfully entangled logics of perception and regulation—to create a historical ontology of environmental sensitivity. I wanted critical discussion about the bureaucratic requirements, laboratory techniques, social norms, observational methods, structural inequalities, statistical models, classification schemes, and other daily practices that made exposure "sensible" at Hanford. And I wanted us to consider the duality that defined the word itself: how *sensibility* wove detection together with reason.[14]

After all, U.S. toxics policy was littered with moments when new sensing techniques had transformed the regulatory landscape. In the 1950s and 1960s, for instance, fallout from aboveground nuclear testing—from bombs fueled by Hanford's own reactors—had fundamentally altered the terms of environmental perception. Radioactive contamination made it possible to visualize industrial impact on a planetary scale, inspiring countless studies about the fate and transport of toxicants worldwide. So too, debates about fallout generated unprecedented levels of funding in the earth sciences, creating new infrastructures around the study of environmental health.

Indeed, Cold War developments in computerized gas chromatography–mass spectrometry revolutionized environmental monitoring, increasing analytical sensitivity by orders of magnitude and revealing chemical residues in air, water, food, homes, and bodies that had previously gone unseen.[15] As scientists at the newly formed U.S. Environmental Protection Agency (EPA) described it, "The identification of pollutants at the part-per-billion level ... has become nearly routine in several EPA laboratories. What was once an impossible task for a staff of 100 working six months sometimes can be accomplished by a skilled individual in a few hours."[16] These enhanced sensitivities (and the hundreds of suspected carcinogens they identified) led to the creation of the Safe Drinking Water Act in 1974.[17]

With new sensory capacities came new legal and regulatory challenges. In 1962, for example, the Food and Drug Administration (FDA) approved the use of carcinogenic diethylstilbestrol in farm animals under the condition that its residues could not be detected in meat after slaughter.

However, as legal scholar William Boyd writes, "By the early 1970s, it had become quite clear that the 'no residue' standard was entirely contingent upon the sensitivity of the analytical technique used to detect residues and that, as such techniques improved, a literal interpretation of the standard—what some referred to as the 'no molecules' approach—would become impossible to meet in the absence of a complete ban."[18] Or, as the FDA's former chief counsel, Peter Hutt, put it, "When the FDA entered the 1970s, the Agency believed that it was feasible to eliminate virtually all carcinogens from the food supply. By the end of the 1970s, the Agency had indisputable proof that this [was] impossible. Thus, it became essential to adjust regulatory policy to accommodate this new scientific information."[19]

In other words, new sensitivities required new sensibilities. Echoing nuclear industry rationales for radiogenic exposure, the FDA and EPA redefined *safety* as *acceptable risk*, making probabilistic reasoning integral to environmental and public health. Within this logic, detection was no longer enough to merit regulatory action. Instead, toxic materials would only be managed if they posed an "unreasonable risk" to human health and the environment.

In addition, the burden of proof fell to regulators rather than industry in key policy moments. The 1976 Toxic Substances Control Act, for example, ordered that the EPA establish unreasonable risk before it could require testing for industrial chemicals (let alone regulate them).[20] This effectively institutionalized ignorance about the majority of such substances on the American market. As the National Research Council reported in 1984, "78% of the chemicals in U.S. commerce with production volume of 1 million pounds or more per year, lacked even 'minimum toxicity information.'"[21]

By adopting nuclear industry frameworks for acceptable risk, EPA significantly expanded the use of cost-benefit analysis in U.S. environmental policy. In fact, this material-discursive shift made risk the very basis for public health legislation, reproducing Cold War notions that economic and environmental protections existed in opposition. "True to its roots in nuclear weapons planning," historian Linda Nash writes, "the paradigm of risk imagined the world as always and unavoidably hazardous, while never asking why."[22] Instead, regulatory infrastructures naturalized these hazards as necessary trade-offs for social progress, making "the seemingly irrational perspective that improvements in human well-being must come at the cost of human health . . . a wholly rational proposition."[23]

One of my favorite short stories comes from Sofia Samatar's collection *Tender*. This fictional piece is told from the perspective of a woman, a "tender," living inside the glass-domed St. Benedict Radioactive Materials Containment Center. Her job, as she describes it, is to monitor toxicity levels within the facility using sensors implanted beneath her skin and to report her findings to the Federal Sustainability Program. These sensors allow her to detect toxicity levels instantaneously, simply by existing in the space. "I can feel the degree of these changes, where they are located, and whether they require observation or action," she tells the reader. "It's like a sixth sense: not at all painful, and far more efficient than collecting and interpreting data. . . .*To tell you the truth, I find it comforting to know how poisonous everything is. I am perfectly attuned to what is good and bad. I always know the right thing to do. Yes, my sensors are strange, but they have given me something akin to a moral compass*" (italics in original).[24]

At first I was drawn to this story for its fictions. After so many years wading through cleanup's arcane bureaucracy, it was satisfying to imagine a world where exposure was simple and administratively legible, its effects measured in real time. I envied the tender her clarity, her perfect toxic attunement, her unambiguous, contaminated comfort.

Yet the more times I read her story, the more parallels I noticed between the tender's telling of nuclear waste management and the official cleanup narratives that I had come to know. Both depended on a fictional boundary between inside and out; both measured impact through an imaginary body. And both reproduced a particular sensory politics of possibility—a formal means for reacting in and through toxic space. Each story tended the relationship between perception and regulation. Each sensed in the service of normative sensibility.

On the last day of WMS I returned to the show floor, weighing the cost of pain against the benefit of watching product demonstrations. This time, Valkyrie was asleep at her post, chest orb unlit, arms hanging limply by her sides. Staring at her now in this more vulnerable state of repose, I found myself imagining that she too had a headache. I wondered about the data she had gathered in the Robotics Pavilion throughout the week, how she had used her own implanted sensors to tend the space. And I thought about the name Valkyrie, which meant "chooser of the slain," a title she shared with the Norse warrior deities who decided life and death on the battlefield.[25] I couldn't help but notice that both the Valkyries of myth and the one sleeping before me were considered guardians in their own right, each one tasked with the double-edged labor of care.[26]

Indeed, care was the primary language spoken in the pavilion—a shared narrative animating mechanical beings and their makers alike. If robotic technologies could replace humans in hazardous work environments, they argued, then the industry's exposure rates would automatically go down. One vendor advertised a collection of tools designed to "eliminate exposure" altogether, promising to keep "operators safe from contaminants."[27] Another made its case in three capitalized words "ROBOTS. REDUCE. DOSE." inviting conference attendees to "drive our robots and select the right [one] for your dose reduction needs."[28] In addition, new sensing, monitoring, and imaging technologies offered enhanced safety possibilities. "Got radiation?" another vendor prompted. "See what you've been missing."[29]

Still, it was difficult to disentangle such promises of protection from the toxic smell in that room. I felt its contradictions in my body, in my own increasingly tender head. I wanted to "see what we'd been missing" about that very moment too. Walking off the show floor for the last time, I remembered something my friend Mark had said years before in the aftermath of an exposure event at Hanford.[30] "DOE told me they didn't find any dimethylmercury in the area," he reported, "but when I asked if they had measured for it, they said no." "And you know what that means," he added with a sarcastic laugh, "if you don't measure it, *it doesn't exist!*"

However, the relationship between measurement and existence is a contingent one, of course: even if you do measure exposure, it exists within the uneven social relations that give it meaning. Just that morning, in fact, I had read yet another news story about the workers who'd inhaled plutonium during a recent demolition project at Hanford. The DOE manager on-site had downplayed the exposures, comparing them to a dose one might receive from a cross-country flight. "He must think we're stupid," one worker told me over lunch before I left for WMS. "That plutonium will be in their bodies for life. It's not the same as a six-hour flight. And it can cause cancer—how could he even say that?!"

It struck me that many of the robots in the pavilion had been created to address other, related failures of nuclear imagination. Weapons production sites like Hanford, after all, were not built to be remediated. There was no end point to the nuclear deterrence model, no vision for a world after the Cold War. As the technical director for Sellafield (England's largest nuclear site) put it, the system simply "wasn't designed for decommissioning."[31] One of cleanup's primary challenges, she continued, was determining how to "get material out of places where it was never designed to be removed from."[32] In other words, how to reckon with a future that nuclear industry never imagined could be.

Such myopia was a frequent topic of conversation at Hanford. My fellow HAB members and I often bemoaned the shortsighted practices of the past, lamenting the multimillennial mess that previous decision-makers had left behind. We liked to imagine ourselves as different: farsighted and responsible, committed stewards of future generations. But sometimes we caught glimpses of our own limited vision. Sometimes, exhausted by cleanup's sensibilities, we admitted that one day our work would be considered myopic, too.

EXPOSURE SCENARIO

Two months after WMS, I traveled to DOE's Argonne National Laboratory for a weeklong training in its RESRAD software. Short for RESidual RADiation, the program used statistical modeling to assess potential exposures at contaminated sites and to derive associated cleanup criteria. RESRAD was designed in the 1980s to be a regulatory tool, our teacher explained, a way to help state and federal agencies address the question: *How clean is clean?* "The answer is always the same," he added with a knowing chuckle, "clean enough to fulfill the regulatory requirements."

In the decades since, RESRAD had become an internationally recognized system for radiological dose assessment. In fact, our teacher said proudly, it was now "the most extensively tested, benchmarked, verified, and validated code in the environmental risk assessment and site cleanup field." Clicking through a series of PowerPoint slides, he told us that RESRAD had been cited in more than a thousand journal articles, downloaded by people in 114 countries, and used to evaluate more than three hundred cleanup sites. In addition, it was the only code specifically mentioned in DOE Order 458.1 as meeting the agency's quality assurance requirements.

Indeed, one need only look to the twenty-eight trainees in the room to recognize the extent of RESRAD's influence. There was a nuclear technology adviser from Stockholm working on a decommissioning project for the Swedish government and an engineer from Ontario designing Canada's Deep Geological Repository. There were scientists from South Korea studying liquid and gaseous wastes and engineers from China interested in emergency response planning. Throughout the week, I got to know safety officers managing mineral sands in Western Australia and private consultants navigating cleanup bureaucracies in Kentucky and Illinois. I heard about radium-rich scale in abandoned oil pipelines in northern Michigan and radioactive archival collections at the Smithsonian in Washington, D.C. I met a hydrogeologist from South Carolina who had worked at Hanford before my time, and we gossiped about the site as if it were a family member.

There was an easy camaraderie among the trainees, a shared paradox of pleasure and anxiety in addressing such toxic challenges. We compared notes about containment strategies and environmental monitoring, commiserated around budget cuts and regulatory fictions. We echoed each other's concerns about radiation protection in deep time. In fact, one scientist confessed that his struggle to manage multimillennial uncertainty was what had drawn him to the workshop in the first place. He needed a crystal ball, he said, half joking over lunch in the cafeteria, and he hoped that RESRAD might be the next best thing.

I was unique among the group as the only person who did not actually plan to use RESRAD after the training ended. I wasn't interested in becoming a practitioner. Instead, I had registered for the workshop because I wanted to spend time in and among its statistical futures. I wanted to understand how RESRAD coded contaminated life, to learn the regulatory math that made cleanup possible.

We began with the basic, conceptual equations. "Big picture," our teacher said, "we are here because there is radioactive contamination in the soil and we need to identify the risks to the individual who interacts with it." "So," he continued, pointing to a cartoon rendering on the screen behind him, "how was this person exposed to this contamination?"

It was a deceptively simple question, he warned. In order to answer it, we had to know something about how the person lived. What daily activities and bodily processes put them at risk from residual radiation? Did they fish? Eat from a garden? Drink milk from local cows? How much time did they spend swimming? What about sleeping? How many breaths did they take per minute, and what did their body actually *do* with the radioactive particles it inhaled?

Second, how much contamination were we talking about, and where was it located in the environment? Was it deep underground or right near the surface? How quickly did it move through the soil? And what were the half-lives of the radionuclides in question? How long before they decayed away entirely?

There was much to consider: precipitation and erosion rates, wind speeds and root depths, leaching models and dose conversion factors. We discussed soil mixing and foliar deposition and water-to-fish transfers. We talked about groundwater flow and bioaccumulative processes. We estimated the amount of milk a farming family might drink in a day. "Now, how would we translate all of this into a cleanup action?" our teacher asked. "That's what we really want to know."

Working in pairs, we turned to a series of word problems designed to put these concepts into practice.

> A site contaminated with uranium-238, cesium-137, and strontium-90 at concentrations of 18.5, 1.85, and 1.85 Bq/g (500, 50, 50 pCi/g) respectively, over a depth of 15 cm is to be cleaned up and converted to a farm. The area of the contamination is 10,000 m^2 (one hectare or about 2.5 acres). The groundwater table lies at 2m below the contamination. Water comes from a well at the edge of the property. Tests show that the Cs-137 and U-238 distribution coefficient is about 15 cm^3/g in all zones.
>
> What is the expected peak dose? When would it occur? What pathways and radionuclides contribute most to this dose? What might have to be done to get the site in compliance with 0.25mSv/yr criteria?[33]

It was slow, detail-oriented work. My desk mate and I plodded through RESRAD's drop-down menus—entering data points, identifying transport parameters, and selecting conversion factors until we reached "peak dose," like a prize, at the end.

The soil was too contaminated for farming in its current state, we concluded. In order to meet federal dose limits for residual radiation, concentration levels would need to be reduced by at least a factor of six. And although the site was currently experiencing peak dose, we anticipated another sustained spike in exposures forty years from now, when contaminated groundwater reached the well at the edge of the property.

There were several cleanup actions that the site could take in order to achieve compliance with federal dose limits. To reduce current exposures, it could add a layer of "clean cover" (uncontaminated soil), lowering risk through additional shielding. Addressing the groundwater issue would require more expensive interventions, however, such as removing the contaminated soil or pumping and treating the contaminated plume. "Given the groundwater problem, perhaps farming is not the most reasonable land-use scenario for this site," our teacher suggested.

Throughout the week, as we moved through increasingly intricate word problems, our teacher returned to this framing again and again. "Ok, now consider your answer," he instructed. "Is this reasonable?" I soon learned that reasonableness was an integral part of the training, a form of rationality written into the code itself. There were some things we simply could not know, our teacher reminded us, and RESRAD was designed to account for that. But the program's users were essential to the rationalization process too. "You have to ask yourself: Are my recommendations realistic?"

At night in my hotel room, between spoonfuls of microwaved soup, I made my way through RESRAD's *User Manual* and *Data Collection Handbook*. Because it was "not possible to predict future conditions with certainty," the manual explained, RESRAD was designed to overestimate dose.[34] Such overestimates served as a protective buffer against uncertainty, like statistical airbags for radioactive space. The trick was to find the right balance between a reasonable overestimate and an outlandish one: to provide a "realistic assessment of future use of the subject property yet be sufficiently protective to ensure that other less likely, but plausible, use scenarios" would not result in excessive dose.[35]

However, the training's frequent emphasis on realism often felt like an exercise in cognitive dissonance to me. RESRAD's world was cartoonish, with homogenous soil columns, perfect rectangles of contamination, and simple, predictable ecosystems. Even the platform's aesthetics—its garish

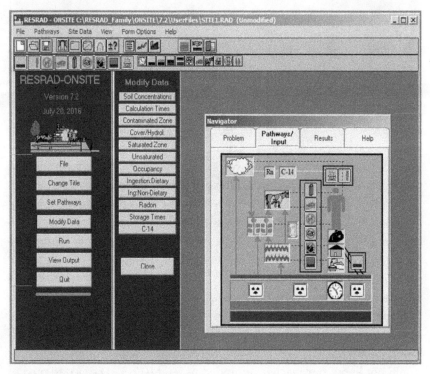

RESRAD navigation page. Image credit: Argonne National Laboratory RESRAD Program.

colors and ancient clip art—seemed designed to highlight the surrealism of our efforts.

Because we had come to the workshop with particular cleanup projects in mind, several trainees expressed frustration with the code's generalizations. There were notable gaps in RESRAD's exposure parameters, omissions that sometimes made it difficult to assess our own contaminated sites. "Remember, this is a *model*," our teacher said in response. "How many sites actually match the assumptions in RESRAD? I don't know of any."

Though I understood his point, I still felt a tug of whiplash with each act of statistical translation. Risk assessment often reminded me of a game of ping-pong, the way it bounced back and forth between general and specific, abstract and concrete, touching down for a split second before hovering in approximation again. Because site managers used

RESRAD to inform remediation planning on the ground, I struggled to reduce its calculations to mere thought experiments. Instead, I wondered how the program materially altered waste management practices—how its assumptions about the future made that future possible.

I was especially interested in the statistical human that inhabited RESRAD's code. Known as "the receptor," this individual embodied an amalgam of behavioral data designed to represent a "realistic but reasonably conservative" life. As DOE's *Data Collection Handbook* explained, its parameters had "been carefully selected," so that "use of these values, in most cases, [would] not result in significant underestimation of the dose or risk."[36]

The receptor drank 1.4 liters of water per day (510 L/year), for example, what DOE considered a reasonably conservative assumption given national averages. Yes, EPA recommended 2.0 liters per day in its own exposure scenarios, DOE acknowledged, but that value was derived from U.S. army data for soldiers in the field and thus was "believed to be an overestimate."[37] And although the *Handbook* stated that 2.0 liters could represent a reasonable "worst-case value," DOE did not find it realistic enough for everyday use.

There were many comparative details like this, each one justified by behavioral studies and probabilistic analyses. Indeed, the *Handbook* often felt less like a description of default parameters and more like an interagency debate about rational living. DOE's receptor tended to eat smaller servings of fish and leafy vegetables.[38] It was also more fastidious in preparing its meals. "The EPA thinks people don't wash their food before they eat it," our teacher told us, laughing and shaking his head.[39]

This was a tension I recognized from countless HAB meetings: that subtle tug of war between EPA and DOE, regulator and regulatee. However, I was struck by the intimacy of these politics in RESRAD—the way that agency imaginations extended to breath and digestion, to the microscale of dust on skin. Even the word *receptor* seemed to instantiate coded sensibilities at a cellular level, as if the statistical human contained tiny, bureaucratic nerve endings. Each data point articulated a distinct sociality of exposure, reproducing regulatory boundaries between "average" and "extreme."[40]

Because cleanup criteria were designed to achieve a particular dose, these daily habits mattered in a practical sense. Washing one's vegetables and eating less fish meant fewer exposure opportunities; 1.4 liters of drinking water produced a smaller potential dose than 2.0. Assigning probabilistic values to behavioral data, therefore, had direct consequences for remediation planning. How the receptor lived informed how the DOE cleaned.

Despite its fine-grained attention to select parameters, however, the *Data Collection Handbook* only offered a pixelated image of contaminated life. There was no sense of the receptor's humanity in its pages, no hint as to how this individual understood its statistical constraints. And though I knew I wouldn't find it, I continued to search for those interior details anyway, longing for greater depth and dimension. What did it *feel* like to measure one's food and water, day in and day out? Where did the mind go when counting particles of dust? Did the receptor experience hypervigilance as burden or comfort in that static, repetitive future?

I knew from my time at Hanford that the receptor was Reference Person, a human template that populated most nuclear exposure scenarios in the United States. Built by committee in 1949, Reference Person began life as Standard Man (later called Reference Man): a young, white twenty-something who lived in a temperate climate and, as the International Commission on Radiation Protection (ICRP) put it, was "Western European or North American in origin, habit, and custom, i.e., caucasoid in type and culture."[41] Though just a rough sketch at the beginning of the Cold War, Reference Man had undergone significant "revision and extension" over the decades, including a major upgrade in 1975.[42]

Documented in a five-hundred-page ICRP report known as Publication 23, that upgrade reimagined Reference Man in breathtaking detail. It described the dimensions of his tongue (7.3 cm from tip to upper edge of the larynx), the specific gravity of his skin (dermis + hypodermis = 2.087), and the precise weight of his fingernails and toenails (3 g, combined). It chronicled the blood content of his brain (503.8 g) and the volume of his snot ("flow ranges from 500 to 1000 ml/day") and the exact length of each of his sperm (54.5 micrometers from head to tail). However, though it reported the shape of Reference Man's heart (a "blunt cone" 9.4–14 cm in diameter), Publication 23 failed to communicate his emotional register. It was difficult to find the person in that ocean of data, to understand who he really was.[43]

On day three of the RESRAD training, I asked our teacher if we could give the receptor a name. I thought it might make for an interesting group activity, a way for us to discuss *whose* life we were building on screen. *Steve drank uranium-238 from the well. Steve failed to wash his leafy greens. At current soil concentrations, Steve would be overexposed.* A few trainees chuckled at the suggestion, and our teacher smiled wanly, assuming that I was either making a joke or a problem.

What I had failed to recognize, of course, was that naming the receptor would undermine his representational value. I was not supposed to be curious about this person's emotional experience. I wasn't supposed to imagine him as a person at all. Indeed, Publication 23 framed Reference Man's anonymity as essential to his utility. "It cannot be stated too firmly that the model values ... represent an *entirely hypothetical* Reference Man not to be related to any particular individual, population, or environmental situation" (italics added).[44] Though he had lungs and hairy arms and intestines filled with exact quantities of gas, he was not a living, breathing human. He was at once meticulously described and unidentified. A very specific no one.

Yet even in his nonexistence, Reference Man embodied powerful assumptions about reasonable harm. There was nothing neutral about the default white male form. Quite the opposite, it was definitional to radiation protection—a deeply consequential condition of possibility. As an adult male, for example, Reference Man was less likely to develop radiogenic cancers than were women and children.[45] Thus, in a regulatory system built around carcinogenic risk, his mere presence had the potential to underrepresent nuclear injury for large portions of the population.[46]

Though Publication 23 also provided an assortment of female parameters, it was up to individual agencies and researchers to decide how (and if) to incorporate them. In 1974, for example, DOE scientists developed a "partially hermaphrodite" model with ovaries and a uterus, using the Reference Man data that ICRP was compiling for its report.[47] Six years later they gave him female breasts, acknowledging the tissue's distinct sensitivity to ionizing radiation.[48] In 1995 they modified his uterus to include a placenta and fetus, allowing it to expand in size over the course of nine months.[49] His body was smaller in this iteration, however, representing both a fifteen-year-old male and an adult female at the same time.[50]

It is worth emphasizing that in designing a dosimetric model for exposure during pregnancy, DOE scientists did not *begin* by creating a separate female form. Instead, they relied on a model "originally developed for a 15-year-old male, with organ masses appropriate for a 15-year-old male" because of the "fortuity" that its general proportions were similar to that of the adult female in Publication 23.[51] These similarities, they argued, permitted the male model "to represent Reference Woman after appropriate changes in masses of breasts, ovaries, and uterus" as well as select values for organs and bones (some of which were "scaled down from the reference adult male" due to a lack of female data).[52]

Such efforts to account for sex-specific risks highlighted the female body's awkward fit within the nuclear industry's male norms. In fact, some computational models did not place female breasts on Reference Man's chest, but instead distributed them as a thin layer throughout his entire body. This meant underestimating exposure to breast tissue from inhaled radionuclides in the lungs and overestimating it for ingested radionuclides in the intestines (where one would not usually expect to find breast tissue).[53] Similarly, as ICRP Publication 130 noted in 2015, "Breast tissue is rarely explicitly defined as a source region in biokinetic models." More often, it is relegated to a more general (and uncertain) "other tissue" category.[54]

It is tempting to seek out a liberatory politics in Reference Man's shifting form—to ask how wearing breasts like a skin might trouble fixed boundaries of sex and gender, or how his pregnancy unsettles heteronormative hegemonies of reproduction.[55] However, to me these models are less about reimagining what the body could be than about maintaining dominant modes of representation. Female data gaps are filled with scaled-down male equivalents; breast tissue is made "other" in the very act of incorporation.

In 2007 ICRP Publication 103 recommended that Reference Man be updated to Reference Person: a "nominal individual" whose anatomical systems and structures comprise an average of male and female data.[56] Though this was an ostensible move toward greater equity, Reference Person continues to prioritize normative, male bodies. Its sex-averaged biology subsumes difference within a broad population-scale frame, masking disproportionate risk to women, children, and other radiosensitive groups.[57]

Publication 103 notes, for example, that individuals with inherited mutations that affect DNA repair (such as those associated with the BRCA, or breast cancer, gene) have a "greater-than-normal sensitivity to the tumorigenic effects of radiation."[58] Nonetheless, it argues that such mutations are "too rare to cause significant distortion of population-based estimates," and thus they need not be considered.[59]

At the same time, Publication 103 touts what it calls the "principle of optimization," recommending that every reasonable effort be made to limit harm by reducing dose rates below official requirements. Still, it clarifies that "reasonable" should be defined by cost-benefit analyses that take "economic and social factors into account."[60] In other words, it explains, "Optimization of protection is not minimization of dose." Rather, it is "the result of an evaluation, which carefully balances the detriment from the exposure and the resources available for the protection of individuals. Thus, the best option is not necessarily the one with the lowest dose."[61]

RESRAD's receptor embodied these logics, optimizing protection via rationalized exposure. The platform's most "reasonably conservative" land-use option, for example, was that of the resident farmer who sourced half of its food and all of its drinking water from contaminated land (the rest of its food came from an off-site grocery store). Reference Person-as-farmer spent more time outdoors than its suburban iteration, exposing it to higher levels of residual radiation. It experienced more groundshine from contaminated soil, inhaled larger quantities of contaminated dust, drank water from a nearby well, and ate fish from an on-site pond. With each activity, its risk of cancer increased, along with the need for more intensive cleanup actions.

The resident farmer was RESRAD's default scenario, our teacher told us, which was one of the program's many conservative assumptions. "We are assuming that everything is happening at these higher levels of contamination," he said. "Is that realistic? No. Then why do it? Because when your dose is higher than realistic, you are being more conservative."

However, by selecting the resident farmer as its most reasonably conservative option, RESRAD was also defining whose life it was possible to protect. The farmer occupied the upper territories of statistical probability, representing the most exposure one might reasonably expect to receive. Anything greater became "unreasonable" by comparison, pushing it beyond the boundaries of regulatory concern.

In 2004 and 2007 respectively, for example, the Confederated Tribes of the Umatilla Indian Reservation (CTUIR) and the Confederated Tribes and Bands of the Yakama Nation created their own exposure scenarios for Hanford's cleanup.[62] They assumed higher breath rates, more drinking water, and traditional foods and medicines, including much larger quantities of fish.[63] In addition, they considered exposure pathways that the resident farmer scenario did not, such as inhaling contaminated steam in a sweat lodge.

These tribes have treaty rights to use Hanford's land in perpetuity, which to my mind makes their scenarios the most reasonably conservative option.[64] So too, Hanford is not unique in its occupation of tribal land; much of the U.S. nuclear fuel cycle, from mining to waste management, takes place in and around Native American communities.[65] Given this, I asked our teacher, why not make a tribal scenario the default in RESRAD?

Though he did not answer my question directly, he noted that users could enter tribal scenario parameters manually, increasing the farmer's risk one data point at a time. However, because the code did not offer a function for steam-based exposures, it could not estimate dose from sweat lodge use. "Know your tools," he advised. "Not every situation is a good fit for RESRAD."

His words reminded me of something an EPA staffer had said years earlier in a presentation to the HAB's River and Plateau Committee. There was a limit to what remediation could do, he told us: "EPA typically cannot protect every individual." Cleanup, he explained, was designed for "sensitive" but not necessarily "hypersensitive" people.[66] "One of the policy goals of the Superfund program, is to protect [the] high-end, but not worst-case" region of the bell curve. "Only potential exposures that are likely to occur will be included in the assessment."[67]

However, the "likeliness" of exposure was informed by the receptor's "normal distribution of sensitivities," meaning that *who* inhabited the agency's model ultimately defined what a "normal" range could be.[68] The same staffer told me, for example, that it was "implausible" to spend the amount of time in a sweat lodge that the tribal scenarios had claimed.[69] Another staffer said that the tribal scenarios' inhalation rates were "not credible" due to faulty assumptions about bodily process. "The Yakama and

Umatilla (CTUIR) have developed their own scenarios, so we run those," he explained. "Unfortunately, they aren't physiologically possible, so we don't choose them. What they did, particularly the Umatilla, is the breathing rate that they chose was from a soldier digging a fox hole, so they were breathing heavy continuously, which you physically can't do. . . . So, for us, we can't choose it because it's not credible."[70]

Later, when I asked a member of one of the tribal scenario design teams about this, he sighed. "That's one thing that they do: they look at the numbers and they say, 'that's not possible.' You know, they look at the fish consumption rates, and they're like, 'no, there's no way that a person could eat that much.'"[71]

The tribes' exposure scenarios could only be taken seriously, it seemed, when they did not challenge settler colonial imaginations of everyday life.[72] Debates about lung capacity and diet, therefore, served as proxies for much larger stakes. Remediation was about more than waste management; it was about who the U.S. government deemed possible to protect.[73]

On the final day of RESRAD, we received signed certificates with decorative gold edges attesting to the fact that we had completed the training. Then we said our goodbyes and boarded the bus that would take us across Argonne's campus for the last time. As we left the facility, I removed the ID badge that I had been wearing around my neck (which an armed guard had checked against my face every morning at the entrance gate) and slipped it into my backpack with the certificate.

I thought about how the week had begun, waiting in line to pick up that badge beside a wall-sized celebration of Argonne's history. "Ever since we were born out of . . . the Manhattan Project in the 1940s," it read, "we have focused our attention on the biggest questions facing humanity." There were black-and-white photographs of scientists bent over their work and images of the first nuclear reactor powering a string of light bulbs. There were descriptions of "intellectual curiosity . . . brought to life" and "breakthroughs that will form a brighter shared future." At the center of the wall was its primary message, the last three words unmissable in boldface text: "Our science **redefines the possible**."

When I look at my RESRAD certificate now, I think about Sofia Samatar's tender. At night in the Containment Center, she reads about "the dawn of the nuclear age . . . moved by the young physicists, their bravery, their zeal." She reads about their regrets too, citing Robert Oppenheimer:

Mural of atomic history at Argonne National Laboratory, 2018. Photo by author.

"The physicists have known sin; and this is a knowledge which they cannot lose."[74] By the end of the story, she is near death, her body literally becoming the waste that she has been tending.

> I sit with the earth as if at the bedside of a sick friend. I am so tender now, I feel the earth's pain all through my body. Often I lie down, pressing my cheek to the dust, and weep. I no longer feel, or even comprehend, the desire for another world, that passion which produces both marvels and monsters, both poisons and cures. . . . I understand that there is no other world. There is only the one we have made.[75]

The tender's story is about care, but it is also about complicity, and her struggle to distinguish between the two. Embodying waste for the Federal Sustainability Program has incrementally conditioned her senses, bounding her capacity to feel, to comprehend, to desire another world. This, more than anything, is what haunts me.

2 Anatomy of a Phantom

At first it was unclear how Case 0102 became contaminated. By the time an employee screening program detected the americium-241 in his urine, it had probably already been there for years. His employer assumed he'd inhaled it, lungs being the most common pathway for the occupationally exposed. Only after his death did postmortem radiochemical analysis offer an alternative explanation. The soft tissues in his left hand were significantly more radioactive than in his right, meaning the americium must have entered unnoticed through a puncture wound.[1]

One week after completing the RESRAD training, I found myself staring at that hand in a long, foam-lined metal box. It was waxy and yellow, stiff fingers outstretched, as if grasping for something just out of reach. Maybe it was the animate quality of that gesture, or the existential angst I brought home from RESRAD. Or maybe it was simply the shock of seeing someone's limb in a carrying case. But as I studied its lively/lifeless form, I felt what I could only describe as grief. Not sadness, per se, but disorientation. The jarring bewilderment of loss.

This reaction surprised me. I did not know 0102 when he was alive, nor was it easy to qualify the appendage itself as human. All that remained of the person was an incomplete collection of radioactive bones encased

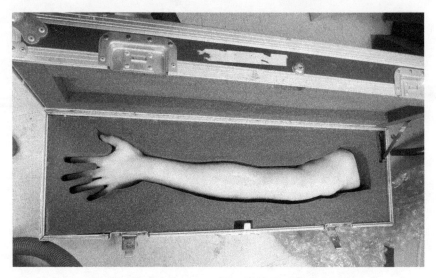

Arm phantom at In Vivo Radiobioassay and Research Facility, 2018. Photo by author.

in "tissue equivalent" polyurethane. In technical terms, it was known as an *anthropomorphic phantom*: a physical model designed to "faithfully mimic" the body and its interactions with radiation.[2] Nuclear science had long relied on such surrogates for experimental purposes. Not only did they permit investigation without exposing live subjects, they facilitated comparative study across institutions and contexts. In fact, 0102's hand and arm had participated in research projects around the world, traveling from one facility to another in the box now open at my feet.

I had come to Hanford's In Vivo Radiobioassay and Research Facility (IVRRF) to spend time with phantoms like this one. I wanted to explore the regulatory relationships between simulation and protection, to understand how representational abstractions haunted nuclear safety. Like RESRAD's receptor, these phantoms translated lived experience into quantitative data, making exposure commensurable with risk assessment. They imagined the body as a rationalized system of interchangeable parts, framing life in statistical terms.

The IVRRF housed Hanford's Internal Dosimetry Program, one of many such facilities across the DOE complex.[3] Built in 1959, it comprised a collection of body-scanning devices designed to measure and assess radiation in living employees.[4] For individuals working with radioactive materials at Hanford, this monitoring was routine. In fact, the IVRRF performed about 7,000 scans per year, including "approximately 5,000 whole body measurements analyzed for fission and activation products and 2,000 lung measurements analyzed for americium and uranium."[5]

When I pulled up to the facility, I realized that I had been passing it for years without registering its existence. The building was nondescript, tucked behind the local credit union in downtown Richland, just one block from the public library where the HAB held its committee meetings. To my left, I could see the high school football stadium with its familiar brick-and-tan façade, the word BOMBERS painted across it in all caps. This, I understood with chagrin, was why I had failed to see the IVRRF despite hundreds of trips down both Knight Street and Goethals Drive. My eye habitually sought out the stadium and its explosive mascot instead, privileging the spectacular over the mundane.

John, a member of the IVRRF staff, met me at the front door and asked me to step onto a sticky sheet of paper before continuing inside.[6] This was to remove dust from the bottom of my shoes, he explained, which could interfere with the facility's scanning devices. "Our goal is to reduce the background radiation in here as much as possible," he told me. "It doesn't remove everything, of course, but it helps."

I thought about the traces I had unwittingly collected on my way to the facility—the particles of home, car, gas station, and coffee shop that now clung to this thin sheet of paper. My own footprints joined others that had arrived before me that morning, creating a collective record of movement through this small pocket of space. I noticed how insubstantial my narrow soles looked beside the thick tread of work boots and wondered what they had tracked in as well.

It seemed appropriate that entering the IVRRF required reckoning with one's dust. The very point of internal dosimetry, after all, was to visualize the invisible. One of the program's primary challenges, John told me, was that they could not simply peer inside the body, nor was it practical to implant dosimetry devices in critical organs (though this had been tried with sheep at Hanford's Experimental Animal Farm during the Cold War).[7] Indeed, the body itself was its own kind of containment device: muscles and fat acted as barriers between internal organs and external scanners; tissues and bones retained unknown quantities of radioactive material. And while it was possible to measure external exposure by placing dosimeters on the outside of the body (John was wearing one himself), the only way to provide exact internal measurements was via cremation and radiochemical analysis of the worker's ashes after their death.

To estimate radioactivity levels within living workers, therefore, the IVRRF used a combination of scanning technologies, statistical models, and anthropomorphic phantoms. The phantoms were a particularly important part of the process, John explained, as they served to calibrate the scanners' detection efficiencies. Designed to represent an "average" worker containing known amounts of radiation, phantoms allowed IVRRF staff to compare actual activity levels in the body with what the facility's scanning devices could detect.[8] These calibration factors then informed dose calculations for Hanford workers, providing context for the measurements that each scan produced. Phantom bodies, in other words, were as essential to the internal dosimetry program as the scanning devices themselves: one technology gathered the data, the other gave it meaning.

The IVRRF kept its phantoms in a series of storage and processing rooms located at the back of the building. John began our tour with the lung and liver collection, opening a tall metal cabinet marked "Caution: Radiologically Controlled Area and Radioactive Materials Area" to reveal about forty phantoms designated by isotope and activity level.

He selected a pair of lungs containing 7.17 nanocuries of europium-152 and held them up for me to see. They were lightweight, he said, made of foamed polyurethane to simulate the physical properties of lung tissue. They were also of high quality, manufactured by DOE's Radiological and Environmental Sciences Laboratory (RESL) in Idaho Falls and meeting its rigorous Laboratory Accreditation Program (DOELAP) standards. Indeed, RESL maintained an entire Phantom Library of accredited organs that it loaned to in vivo monitoring facilities across the country.[9] Whenever he needed a part, John said, he just called them up and ordered one.

Beside the cabinet was a torso phantom with a removable chest plate and a stack of "tissue-substitute chest overlays." John lifted the plate, flipping it over to show me the phantom's rib cage, which had been fashioned with "artificial bone-substitute material."[10] Effective tissue simulation was an essential part of phantom construction, he explained, as it affected

Lung and liver phantoms at In Vivo Radiobioassay and Research Facility, 2018. Photo by author.

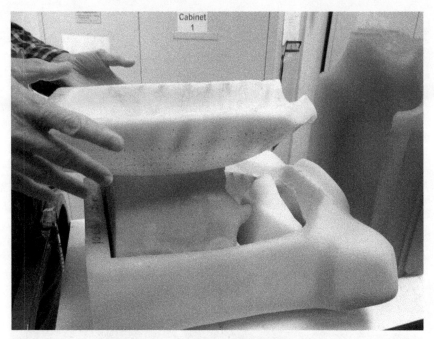

Lawrence Livermore National Laboratory (LLNL) Realistic Torso Phantom at In Vivo Radiobioassay and Research Facility, 2018. Photo by author.

radiation attenuation and detection efficiency. Scientists had been investigating tissue equivalence since the early 1900s, debating the relative merits of water, wood, wax, rice flour, plastic, rubber, Styrofoam, plexiglass, oatmeal, and many other mediums over the years.[11]

This variety in substance and form had made it difficult to compare calibration factors across DOE facilities in previous decades. By the early 1970s, in fact, DOE's Intercalibration Committee for Low-Photon Energy Measurements had determined that "most of [its] laboratories underestimated the amount of activity by as much as a factor of 3 or more," making their dose calculations "grossly unreliable."[12]

In response, the committee designed the Lawrence Livermore National Laboratory (LLNL) Realistic Torso Phantom to "intercalibrate detector systems" across the DOE complex.[13] Modeled from a cadaver that approximated Reference Man from the University of California San Francisco's anatomy department, its polyurethane form contained an artificial rib cage, a collection of "simulated organs and spacer blocks," and three tissue-equivalent overlays to accommodate a variety of chest-wall thicknesses: 100 percent Reference Man muscle, 50 percent–50 percent Reference Man fat to muscle, and 87 percent–13 percent Reference Man fat to muscle respectively.[14] DOE made sixteen exact copies of the phantom and distributed them to its laboratories around the country, where they became integral to dosimetry science and practice.

The committee had designed the Livermore phantom to be "rugged enough" for routine laboratory use, and it certainly appeared that way to me.[15] After nearly half a century of labor, it was scuffed but solid, its surfaces brushed with the patina of age. John showed me where the radioactive lungs fit within the chest cavity, completing a 3-D puzzle of organs and spacer blocks. Then he lifted the entire torso so that I could see it from behind, pointing to an area near the base. "See that part of the low back where it is unnaturally flat?" he asked. "That's where the cadaver melted when they were making it."

I hadn't noticed this detail, but once John pointed it out, I couldn't see anything else. It felt like the most human part of the being before me, a reminder that this body had once been *a body*. The phantom's unnaturally flat back recalled the natural plasticity of its source, and the messy, imperfect process of replication. By highlighting it, John seemed to be breaking the fourth wall, refusing the boundary between story and audience. I saw both the originary form and its phantom script, the representative human at once solid and melting.

In that spirit, I asked John how this male phantom played the role of female employee. I was particularly interested in his breasts: Where were they? What were they made of? Did RESL make attachments in a variety of shapes and sizes? I had read about the ways that breasts reduced detection efficiencies, especially when measuring radioactivity in the lungs. Not only did detectors tend to be positioned farther from female chests, but the relative shape, size, and glandularity of breasts affected lung counts as well.[16] In fact, one recent study had found that "for bigger busts and breast sizes, the detection efficiency showed to be up to ten times lower than the ones measured with the Livermore male torso phantom."[17]

Because breasts posed a unique dosimetric challenge, I was excited to hear how the IVRRF had managed them. I imagined another tall metal

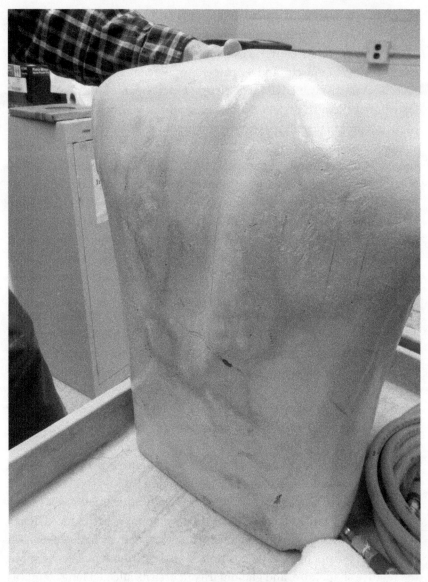

Lawrence Livermore National Laboratory (LLNL) Realistic Torso Phantom at In Vivo Radiobioassay and Research Facility, 2018. Photo by author.

cabinet, its shelves filled with breast phantoms ranging from large to small. I wanted to know how they had simulated glandular tissue, a matter of debate in mammography circles (I had read multiple studies suggesting beeswax, for example).[18] And I wondered about the angle of repose the IVRRF's phantoms assumed, as gravity changed breast shape and thus detection efficiency.[19]

Rather than launching into a detailed explanation, however, John simply replied, "We like to think of this body as unisex." He went on to describe how one *could* create a pair of makeshift breasts by filling bags with gel and placing them on the phantom's chest, but the IVRRF did not do this for routine counting.[20] Instead, it used the worker's height and weight to address bodily variety. And while he acknowledged that large breasts had the potential to influence dose calculations, John did not agree that the difference would be significant. "In general," he said, "we are making decent estimates." In fact, the IVRRF had recently undergone its DOELAP review (required every three years) and passed with flying colors.[21]

However, as far as I could tell, DOELAP conferred its accreditation based on tests conducted with phantoms that did not have female breasts.[22] This appeared true across the DOE complex where, I would later learn, there was a "total absence of female calibration phantoms."[23] One article explained that "currently, there are no physical female phantoms available because such models are difficult, time consuming, and expensive to fabricate."[24] But to me this absence was about more than technical difficulty and time constraints; it also spoke to a broader politics of representation in nuclear industry. John had pointed out multiple flaws in the Livermore phantom: in addition to its unnaturally flat lower back, it lacked scapulae and spine, which produced uncertainties when used from behind.[25] Yet he had failed to address the phantom's missing breasts until I brought them up. They seemed irrelevant to the task at hand.

In this, John echoed a longer history in medical science of treating the female body as adjunct to the male exemplar.[26] His suggestion that one could simulate breasts by placing bags of gel on the phantom's chest reminded me of how DOE had affixed them to its own Reference Man model in 1980.[27] In each case, adding female breasts did not unsettle hegemonic sex and gender categories; it reproduced them, positioning one form as outlier, the other as standard.[28]

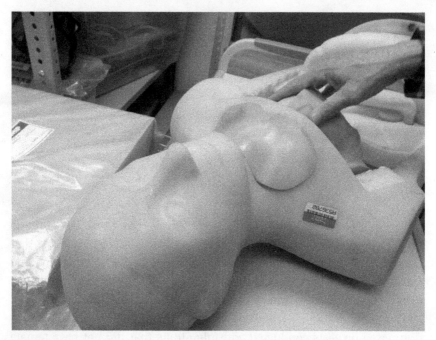

Fission-Product Phantom at In Vivo Radiobioassay and Research Facility, 2018. Photo by author.

Near the end of my tour, we stopped by a storage room filled with an assortment of body parts and other tools: a pair of torsos beside a shop vac and a pile of used bubble wrap, a rolling cart stacked with arms and legs, an array of cardboard boxes and thyroid phantoms lining the back wall. One of the torsos was known as the Fission-Product Phantom, an iteration of the Livermore phantom with more anatomical detail. Created to "meet the needs of nuclear power stations and other facilities that require more organs," it included a heart, thyroid, kidneys, stomach, pancreas, spleen, and small and large intestines.[29] It also came with a head and neck (and a "neck cover plate for thyroid access"), as well as arms and legs that articulated at the hips, elbows, and knees.[30]

These additional features were supposed to make the phantom more realistic, but I couldn't get past the surrealism of the being itself: its monochromatic organs and blank stare, its detachable limbs, the way its neck cover plate didn't quite fit. Phantom care felt like an epistemological seesaw to me, an exercise in continually shifting registers. As John lifted the surrogate's chest plate and removed its heart, I felt myself stumbling between human and object, fragment and whole, possible and impossible.

Though it was easy to tell the difference between a living body and its referent, I kept tripping over the implications of their relationship. What were the transitive properties of this pairing? When and how did one body become the other? I could hear my RESRAD teacher warning, *Remember, this is a model*, as he had when trainees noted the platform's omissions and imperfections. *How many sites actually match the assumptions in RESRAD?* he had said. *I don't know of any.* He wanted us to understand that perfect representation was not only impossible, it had never, in fact, been the goal. Quite the opposite: abstraction was necessary, even foundational, to a system designed for statistical people.

Standing beside the Fission-Product Phantom, I thought about something former EPA policy adviser Lisa Heinzerling had said at a conference about ethics and health. "There are no identified lives in U.S. environmental law," she told the audience. "They are *all* statistical lives. They are statistical lives before we step in to protect them and they are statistical lives after they are helped or after they are hurt."[31] The statistical person was a product of Cold War regulatory initiatives, a "new kind of entity" created to legitimize risk assessment as a form of bureaucratic care.[32] "A primary feature of the statistical person is that she is unidentified," Heinzerling explained, "she is no one's sister, or daughter, or mother. Indeed, in one conception, the statistical person is not a person at all, but rather only a collection of risks."[33]

John turned to the thyroid phantoms, an assortment of headed and beheaded necks along the back wall, and selected an amber-colored thyroid insert from a plastic bag. He compared its shape to a butterfly, another imperfect representation, and explained that it came in a variety of radioactivity levels. Like the lung and liver phantoms, these thyroids were standardized, designed to fit multiple simulated bodies, and thus offered a wider range of applications. However, not all phantoms at the IVRRF were interchangeable in this way, he said. A few had been fashioned from the bones of an exposed worker, making them inherently unique.

It was then that John opened the long metal box containing Case 0102's left hand. And it was then, as I stared at those outstretched fingers, that

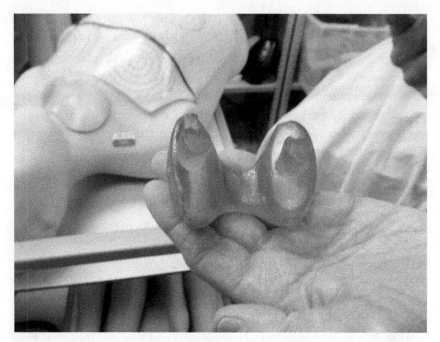

Thyroid phantom at In Vivo Radiobioassay and Research Facility, 2018. Photo by author.

I felt a sudden wave of grief. Children's author Lemony Snicket writes that grief is "like walking up the stairs to your bedroom in the dark and thinking there is one more stair than there is. Your foot falls down, through the air, and there is a sickly moment of dark surprise as you try and readjust the way you thought of things."[34] I have yet to find a better description. It was what I had felt when my parents died years before, and it was what I felt in that storage room, standing beside 0102's hand: a jolt of confusion, a jarring (ab)normality. *Human-object, fragment-whole, possible-impossible.*

Ever since I had seen my first phantom in a fallout archive years earlier, I had struggled to make sense of the name. Why had nuclear science adopted this term? Why not *model* or *manikin* or *dummy*? While each dealt with simulation in some shape or form, the word *phantom* had a more fantastical quality. It meant illusion and unreality, "something merely imagined," the need to scratch an amputated limb.[35]

But the IVRRF's phantoms were more than artifice; they were physical beings that made meaning in relation. They worked in concert with the living, translating contaminated bodies into statistical forms that could receive abstract care. The phantom's labor was distinctly material in this sense, a specifically calculated haunting. It made the invisible visible while also conditioning what it meant to see.

Case 0102's hand embodied a feeling I had carried with me throughout my tour of the IVRRF. I saw in his reach a desire to bridge the representational gap, to reconcile the model with the life that it contained. I was moved by both the effort and its impossibility, grieving what it had made and unmade. That imperfect reckoning, I realized, was what gave the phantom its definition.

HUMAN PRODUCT

Case 0102 donated his body to the U.S. Transuranium Registry (USTR) in 1979. His was the first whole body donation the USTR had received, and it was met with both gratitude and excitement. Hanford had been collecting bone, lung, and liver samples since 1949 in a "modest program of postmortem tissue sampling" called the Hanford Autopsy Study.[36] In its early years the study focused on plutonium, comparing activity levels in workers' tissues after their death with urine samples from their life. It sought to identify how much product left the body through excreta versus how much remained in the organs and bones. Then it used that data to refine the site's biokinetic models, improving estimates for permissible dose.

When study authors presented their data publicly for the first time in 1967, they reported that "low, measurable" amounts of plutonium had been found in both Hanford workers and area residents.[37] However, they attributed the majority of that material to "fallout from nuclear weapons tests rather than occupational exposures or environmental releases from the Hanford site."[38] By framing it as a distributed condition rather than the work of a particular place, the study effectively decoupled fallout from its origins. It dismissed the Hanford-made plutonium that had fueled those nuclear tests and then returned, soon after, on the wind.

Study authors also made the case that more data was needed, advocating for a nationwide program of research. In response, the Atomic Energy Commission (AEC) established the National Plutonium Registry and began recruiting workers from across the weapons complex. Registrants could elect to participate at a variety of levels, from authorizing an autopsy to donating part or all of their bodies to science. In 1970 the program changed its name to the USTR, recognizing a broader range of radionuclides, and in 1992 it merged with another program to become the U.S. Transuranium and Uranium Registries (USTUR). That same year, it expanded again to house the National Human Radiobiology Tissue Repository (NHRTR), an archive of bodily materials collected at autopsy.[39]

I visited the USTUR in early August 2017, during one of the worst wildfire seasons in Washington's recorded history. The sky above Hanford was rust colored, the air sickening. People on both sides of the Cascades had itchy lungs and red-rimmed eyes. As I drove to the facility, the song I had been half-listening to on the radio ended and a DJ reminded people to stay indoors for their protection. "Watch out for Smokezilla!" he said with gusto, and I marveled at how something could be at once cartoonish and frightening.[40]

It wasn't just the fires. For the past week, president Donald Trump and North Korean leader Kim Jong-un had been threatening each other with nuclear apocalypse. Kim had pledged "thousands-fold revenge" for U.S. sanctions, and Trump had promised "fire and fury like the world has never seen"; both were boasting that they had the power to end life as we knew it.[41] There were reporters in Guam interviewing residents about the possibility of being caught in the crossfire and pundits in Washington, D.C., recycling well-worn debates: *See, this is why we need to increase funding for nuclear weapons programs! No, this is why we need to rid the world of these weapons once and for all!* They performed their familiar choreography—back and forth, round and round—until the stalemate was its own form of catastrophe.[42]

The USTUR was out by the Richland airport on Terminal Drive, a light industrial area southeast of the Hanford site. I met Alex at their administrative offices, where we chatted for several hours before walking to laboratory facilities nearby.[43] Alex was loud and energetic, abundant with information and enthusiasm. He clearly loved his work and did it with care.

And it was my curiosity about how care was imagined in and through places like the USTUR that had originally inspired my visit. If cleanup was designed for statistical people, then I wanted to know how such beings came to be. I meant this literally: What technoscientific practices took the exposed body apart and then reassembled it into the data that informed nuclear safety? I wanted to see that distillation process up close, to understand how someone could be deconstructed and made whole again in abstraction.[44]

When we entered the cavernous room at the center of the laboratory building, I saw that it was filled with freezers. There were about twenty of them: long, white, and waist high, each one marked with a bright orange biohazard sticker. Pulling a small point-and-shoot camera from my bag, I snapped a quick photograph of the room and then felt immediately uncomfortable about it. If the freezers were full of people, as I assumed, I was aware that I hadn't asked their permission. I did not know these workers. I had not spent time with them or heard their stories. My own documentary efforts felt like another form of exposure.

Together, we walked to a series of tall metal shelves stacked with plastic shipping boxes. This was part of the NHRTR, Alex explained, which made materials available to scientists around the world. On the front of each box was a list, detailing its contents by case number, sample number, and tissue type: lymphatic, skeletal vertebrae, dura matter, rib, femur, testis. A bar code in the top left-hand corner provided further information when scanned.

Gesturing to a nearby table on which were absorbent packing sheets, rubber bands, and paraffin wax, Alex described the upgrades staff were making to the archive. Most of the boxes contained bottles of digested tissue: pieces of organ and bone that had been dissolved in acidic solution. The archive's acidity had proven a practical challenge over the years, as vapor seeping from the bottles degraded the containers they once used. Pointing to a hole in an old cardboard box on a top shelf for emphasis, he said that the plastic boxes cost more, but it was worth it. These samples were a critical scientific resource, he told me, and they were nonrenewable. "You can't restock these materials. When they are gone, that's it."

I was simultaneously fascinated and unsettled by this clean, organized space. Nearly everything about it felt commonplace: the metal shelves and fluorescent lights, the linoleum floor and Ziploc bag of rubber bands. If it weren't for the itemized lists of body parts on the boxes before me, I could have been in any industrial warehouse. Disfigurement had been so effectively rendered that it would be easy to forget that I was standing among the dead.

In many ways, this was intentional. Alex told me that donors were immediately "de-identified" upon arrival, recognized only by their assigned case number. De-identification protected the individual's privacy, but it was also critical to the body's rationalization. As a worker was processed—autopsied and disassembled, then turned to ash and dissolved in acid—their case number acted as an originary trace, affixed to the bags, bottles, and boxes they came to occupy. It thus performed a quantitative codeswitch essential to the rationalization process, "disenchanting" the body in the Weberian sense and allowing life to become data.[45]

Importantly, however, for this codeswitch to be effective, it could never be complete. Without the original, lived body as referent, its constituent parts would have less scientific value. The USTUR's mission was to better understand the biokinetics of exposure: how transuranic elements like plutonium, uranium, and americium were metabolized throughout the course of a person's lifetime. Program scientists analyzed the body's relationship with nuclear materials, tracking radiogenic movement from breath to bloodstream, organ to bone. By studying the embodied conditions of occupational exposure, they sought to improve the science of internal dosimetry and to develop better models for dose assessment. This research, Alex said, could have wide-ranging applications in radiological protection, risk calculation, and workers' compensation.

The USTUR's data was distinctive, he continued, because its scientists had the privilege of taking the body apart. Rather than estimating exposure by extrapolating from a living worker's urine or dosimetry scans, they performed more directed, microlevel investigations. Using radiochemical analysis, they could measure exactly how much plutonium remained in a worker's lungs and how much had moved to their liver. They could observe whether americium became uniformly distributed throughout the skeleton or if it persisted unevenly in a particular extremity. Such data could then be used to inform statistical models for assessing contamination in living workers as well.

While I was uneasy among the boxes, I was also moved by Alex's genuine respect for their contents. Maintaining this archive was a moral obligation, he told me, not only because it contained critical scientific data, but also because these people had placed their bodies in the USTUR's care. Alex honored their wishes as if they were his own. "I think about how I would feel if the donor was my father," he said.

Processing registrants, therefore, seemed to require a kind of double labor: knitting tissues to their lived histories via numerical code while disassembling the individual into an anonymous list of parts. Indeed, the USTUR's website presented donors as raw substance, visualizing their bodies as a collective pie chart of "available material": skeletal (48%), muscle + skin + fat (22%), glands (3%), circulatory (4%), and so on.[46] It also depicted the archive as a word cloud map of anatomical terms in the shape of the United States: *lymph nodes* straddling the Rockies, *scapula* and *rectum* lining the Great Lakes, *skull* and *vertebrae* nestled into the Gulf Coast.[47]

When I visited the USTUR's website for the first time, I wasn't sure what to make of that map. What message was it trying to communicate? Was this simply a reminder that registrants came from across the country, or was it providing a deeper social commentary about structural exposure?

It was only later, while reading slides from a conference presentation by program staff, that I considered another possible meaning. Entitled "USTUR Research: Land of Opportunity," the slides described a variety of donors within the registries' archive. There were sixteen members of Los Alamos's UPPU (You Pee Pu) Club, who had worked on the nation's first nuclear bomb, and Case 0246 (Hanford's "Atomic Man"), who had received the highest recorded americium exposure in U.S. history.[48] There were two whole-body cases with "long-term retention [of plutonium] in the upper airways" and sixteen partial-body cases containing "high-fired PuO_2 aerosols." There were donors with uranium and beryllium, zirconium and thorium, curium-244 and neptunium-237. The final slide contained the pie chart and word cloud map I recognized from the website, along with the phrase, "So many opportunities—So little time!"[49]

Contaminated workers were an abundant natural asset in this frame. Their exposures gave the map its shape, performing the material and metaphorical labor of nation building. This cartography evoked Cold War narratives that, as anthropologist Joseph Masco has argued, "transformed the apocalypse not only into a techno-scientific project and a geopolitical paradigm, but also into a powerful new domestic political resource."[50] By calling the tissue archive a "land of opportunity," the USTUR seemed to reimagine radioactive bodies as social-scientific goods that advanced the broader nuclear-national project.[51]

I followed Alex through each stage of the datafication process: from the autopsy room to the freezers to the ashing room to the radiochemistry laboratory and, finally, to the counting room, where an alpha spectrometer produced line graphs of plutonium on a computer screen. As we moved from place to place, I saw the registrants in double vision, two images simultaneous and overlapping. The first refracted their bodies into ever smaller pieces (literally becoming dust), until I struggled to remember that they were once whole. The second concentrated them into their most radioactive form, distilling and refining until they *became* the line graphs on screen.

Affixed to the wall beside the computer was a laminated poster with the words *Remember Personal Accountability* and a list of individuals who had secretly been injected with plutonium during the Manhattan Project and early Cold War.[52] I had read about these covert medical experiments, which exposed unknowing subjects to radiation for the sole purpose of studying its effects. Some had been code-named for the universities where they received their injections (CAL for UC Berkeley and CHI for University of Chicago), and others bore the more ambiguous acronym HP (these were based at the University of Rochester).[53] Alex asked if I knew what HP stood for and I incorrectly guessed Human Plutonium. "Close," he said, "but what was the code word for plutonium in the 1940s? Product."

Human Product. The title felt strangely apt. It wasn't just that humans literally embodied the bomb; it was that humanity itself had become an infrastructure of nuclear development. As historian M. Murphy writes, the Cold War manufactured more than weapons and waste; it also bound living-being to the economy at the scale of the population. This distinct "economization of life" conjured humankind as an "experimental object," calibrated in and through its "ability to foster the macro-economy of the nation-state."[54] Aggregate life in the twentieth century was a historically specific epistemic frame that reproduced uneven categories of human worth. Its technoscientific practices "designated both valuable and unvaluable human lives: lives worth living, lives worth not dying, lives worthy of investment, and lives not worth being born."[55]

The statistical person in environmental law is a human product of that history. According to Lisa Heinzerling, a statistical life is "a life expected to be lost as a function of probabilities of death applied to a population."[56] Statistical people are "explicitly priced in advance of their deaths," afforded differential value according to "age, health, disability status, and wealth."[57] Environmental regulation, therefore, involves "a pre-killing weighing of the choice to kill ... economic costs of pollution-reducing strategies are balanced against the value, stated in terms of dollars, of the people whom the environmental hazards in question will kill."[58] This "knowing killing" is refined and abstracted through the risk assessment process, optimizing life and death in the service of the broader national economy.[59]

Critically, economization relies on aggregate life as both its epistemic logic and bureaucratic instrument. Macrological scale allows statistical people to absorb impact in fractionated units like *micromorts* and *microlives*: probabilistic "chances" distributed among "the population" as a whole.[60] Risk reduction is often framed as a financial transaction, a specific sum traded for a specific percentage of avoided death or prolonged life.[61] For economist Kip Viscusi, the "pertinent value" in such regulatory interventions is the individual's "willingness to pay" for adjustments in life expectancy. "What we are purchasing with our tax dollars is not the certainty of survival," he writes. "Rather, it is the incremental reduction in the probability of an adverse outcome that might otherwise have affected some random member of our community."[62]

Viscusi's argument betrays a familiar conceit in environmental risk assessment that "adverse outcomes" are "randomly" borne. "Our community" does not represent collective experience in this frame, but rather an accumulation of indiscriminate effects. It averages exposure across the population while individualizing responsibility through a "willingness to pay" for its abatement.

Statistical people, in other words, mask social inequities even as they bear them unevenly. Willingness to pay implicitly assumes *ability* to pay, granting the wealthy greater access to risk reduction. So too, metrics like "quality-adjusted life-years" (QALYs), which calculate the percentage of a statistical life saved by an environmental regulation, privilege the young, healthy, and able-bodied who have more years left to live. And because life expectancy is strongly associated with social relations like race and socioeconomic status, QALYs may assign disparate value according to the amount of statistical life one might "expect" in the first place.[63]

Cold War technoscience defined accountability through these and other contingent forms of counting. Environmental protections sought to make risk generalizable, untangling exposure-related illnesses from their lived contexts while at the same time reproducing hierarchies of human value in the name of the nation's economic health. By centering the statistical person, regulatory infrastructures made risk the subject of intervention and de-identification a condition of its care. However, as Heinzerling argues, "Simply calling what we are valuing 'risk' (or 'micromorts') . . . does not change the fact that real lives, not statistical lives hang in the balance. . . . If a person dies due to an environmental hazard, a real person dies . . . our inability to identify that person by name does not change the fact that she has died."[64]

The *Remember Personal Accountability* poster evoked the doubled-ness I had felt throughout my tour of the laboratory: how the donors were at once "available material" and somebody's father, de-identified tissues and members of the UPPU Club.[65] The call to remember in the counting room, where exposed bodies became data, felt difficult to ignore. Remembering, after all, is the opposite of dismembering. It means "to put together again."[66]

By the time I left the USTUR, it was early evening. Alex had given me a bright red T-shirt with the program's logo as a parting gift, and I tossed it in the backseat of my car before heading west toward Hanford.[67] The smoke still hung heavy and orange over the site, erasing the reactors and canyon buildings that usually marked my progress home. I looked for them anyway out of habit, squinting through the haze for evidence of their shadowy forms.

As I approached the B reactor, which had fueled the world's first atomic bomb, a DJ on the radio joked that nuclear war was a great way to avoid household chores. "I told my wife: 'What's the point of taking out the garbage if we are all going to be dead tomorrow?'" His words reminded me of another fatalistic quip I had heard that anxious week, borrowed from a

1952 film called *The Atomic City*. It asked simply, "What do you want to be *if* you grow up?," making life conditional with a single word.[68]

RE-MEMBERING

Case 0102 was a doctoral student at the University of California Berkeley in the 1950s and then a bespectacled radiochemist at LLNL. He was a father of three. A long-distance runner and biker. He loved backpacking among the granite peaks of the Sierra Nevada, often wandering off trail for days at a time. I began looking for these and other details immediately following my visit to the IVRRF, unable to shake my grief at seeing 0102's outstretched hand. The feeling was sticky. It persisted for weeks like a residue, insisting that I identify who he was.

First, I found him in a 1985 special issue of *Health Physics* dedicated to his postmortem analysis at the USTUR. It described his graduate work under Manhattan Project scientist Burris Cunningham and his death from malignant melanoma at age forty-nine.[69] It speculated about the moment, twenty-five years earlier, when he had unwittingly absorbed americium-241 through a puncture wound in his left hand. But mostly, it chronicled his transmutation from body to data: the layers of precision and uncertainty that attended the research process, the knowing and unknowing his donation had made possible.

Study authors recounted a challenging exercise in deconstruction and interpretation, at times marked by extenuating circumstances. There was the airline strike that delayed 0102 in transit for more than a week, causing his brain and internal organs (which had been packed in dry ice) to thaw and desiccate. There was the power failure at a Seattle mortuary where he rested on his way to Hanford, leaving him unrefrigerated and decomposing for days.[70] There was the shattered beaker of digested tissue that could not be recovered, erasing radiochemical data for his left thigh. And there was the USTUR freezer that had inadvertently been wired to a light switch (turned on and off as researchers came and went), in which repeated freezing and thawing dried out his bones.[71]

Despite these and other caveats, 0102's contributions to the field of metabolic modeling were clear. As the first whole-body analysis, he provided new information about americium's biokinetic properties, especially its intraskeletal distribution.[72] So too, his radioactive bones produced four anthropomorphic phantoms—head, torso, left arm, and left leg—that would go on to inform dosimetric research around the world.[73]

Though it was relatively easy to find 0102 in death, however, I struggled to locate him in life. My search led me through old newspaper articles about the lawsuit his family had filed against the University of California Regents (related to his exposures in graduate school) and a 1980 report about melanoma rates among LLNL employees.[74] Several articles included the name of 0102's wife, which I eventually (after many dead ends) traced to an email address for what I thought might belong to one of his sons.

At first I was so excited by the lead that I failed to consider what it would actually mean to contact a member of 0102's family. It wasn't until I sat down to write the email that I realized I had no idea what to say. Where to begin such a message? *Hi, you don't know me, but I recently saw your father's phantom hand in a box and I can't stop thinking about it. Want to chat?* I struggled to articulate my motivations, not only to this stranger but to myself as well. Why did I continue to feel so haunted by that hand? And in reaching out to 0102's family, what was I hoping to achieve?

My search was part intellectual curiosity and part desire to refuse the anonymity that made exposure reasonable in environmental policy.[75] Yet as I stumbled through that email, I had to admit that my motivations were personal too. Like 0102, my parents had died of cancer at a young age, and my own body bore the scars of the disease. Navigating the territories of grief and survivorship had left me feeling adrift, lost in the carcinogenic crowd. I wanted more than Viscusi's "random" impacts and indifferent probabilities. I wanted more than a statistical life.

To my surprise, 0102's son responded to my email less than thirty minutes after I sent it and agreed to talk with me on the phone. Alan Gunn, it turned out, lived in the tiny northern California town of Arcata, where he worked as a property manager for a rental company. I had lived in Arcata twenty years earlier, and we chatted about the area, laughing about small worlds. I asked him about neighborhood businesses I remembered, and when I mentioned my first apartment, he laughed again, saying he managed the complex and knew it well. Then he paused for a moment as if lost in thought and said, "You know what? I think I rented you that place!"⁷⁶

My mind traveled back to the overcast afternoon when I had moved into the apartment. I was eighteen years old, my childhood bed crammed into the back of a borrowed car along with books, clothes, and slippery stacks of CDs. I remembered thrilling at the novelty of rental-related tasks: setting up my first accounts with the electricity and telephone companies, scheduling a time to get keys from the property manager and pay him the deposit I had spent the summer earning.

My memory of that day didn't include Alan, but he must have been there to let me in. He must have given me a tour of the apartment and handed me the keys. All I remembered was the glow of pride I felt walking through the front door, how I loved its shabby details because they were mine. It was an important moment in my life. A marker of transition into adulthood. I couldn't believe that 0102's son had been part of it!

Like me, Alan had come to Arcata to attend Humboldt State University. Also like me, he had majored in geography, another connection that took us by surprise. Though he was a generation older, we had even shared a professor and taken the same classes in Founders Hall.

When I began looking for 0102, I hadn't expected to find so many kindred experiences. Alan and I had walked the same forested trails behind campus, shopped at the same local grocery store, and listened to the same professor talk about Nicaraguan politics. He had rented me my first home of my own. It wasn't a distant connection, but an entangled set of stories and spaces. Not six degrees of separation, but one.

It was more than shared geography, however. In addition to the specific and unexpected ways our paths had crossed, we had both lost our parents to cancer early in life.⁷⁷ Alan had just graduated from high school when his father passed away. I had recently finished college. His father died four months after diagnosis. Mine died after three. We talked about the shock of that sudden absence and the tenderness of our young grief. We talked about what it had meant to live with the loss.

And we talked about 0102, whose name was Stuart R. Gunn, Alan's voice warm as he recalled the details of his father's life. Stu had grown up in New England and attended the University of Massachusetts Amherst before moving west for his doctoral program. He was an avid skier and

Stuart R. Gunn in the High Sierra, 1974. Photo by Alan Gunn.

bicyclist, a skilled mountaineer. He ran marathons, including one just four months before he died. Alan and his father often went hiking together, an activity that brought them closer in his teenage years. "Dad taught me to love the mountains," he said, and it was a feeling that had only deepened with time.

A few minutes after we hung up, Alan sent me photographs from one of their many backpacking trips in the High Sierra. There was Stu cooking dinner on a small camp stove. Stu walking a dusty cliffside trail. Stu gazing out across a sea of granite beneath a deep blue California sky. "Now he has a face," Alan wrote, "and is not just a number or a bunch of old bones."[78]

I was grateful to see Stuart Gunn in the flesh. It was a relief to fill in his details, to note his wiry frame and tanned skin, to find his left arm still swinging by his side. I had located the person who haunted the simulation and, in some small way, the act felt healing. Yet I also felt its incompleteness, aware that my own reconstructive efforts were full of erasure too. If re-membering was reparative, it was not the same as closure.

It wasn't until I finished writing this chapter that I realized closure had never truly been my goal. I wanted more than the disembodied limb, more than the archive of de-identified tissues, more than the calculated lives that inhabited cleanup's future. But I also wanted to recognize the limits of representation in a statistically contaminated world. I wanted to identify Stu without ignoring the powerful conditions that had rendered him statistical in the first place.[79]

Sociologist Avery Gordon argues that "to write stories concerning exclusions and invisibilities is to write ghost stories."[80] It is to understand haunting as a material tension between the seen and unseen, to investigate the social structures that make some lives more legible, more grievable, than others.[81] Further, it is to analyze the tactical, both/and qualities of phantom reckoning, to feel the twin pulls of the abstract and concrete.

I have struggled with this. At times, searching for Stu felt like a rescue mission, an effort to deliver him (myself) from abstraction. Yet in my urgency to locate his originary form, I often minimized the afterlife he continues to lead.[82] In dosimetry science, for example, 0102 is a distinctly embodied figure valued for his anonymous yet specific history of exposure. Study authors describe the unique realism of his "naturally-contaminated" bones, comparing them with the lesser "artificially-contaminated ones" available on the market.[83] In fact, Stu-as-0102 is frequently the most concrete being in the room: the "actual exposure case" versus the "fabricated" one.[84] His phantom presence, in other words, resists the disembodiment of his commercially manufactured kin.

However, these same studies also note the imperfections and "disadvantages" of Stu-0102's constructed parts.[85] They describe the two halved skulls (only one of which is Stu's) glued together to create his phantom head and the air bubbles sprinkled throughout his tissue-equivalent brain.[86] They mention the sternum, pinky toe, and ring finger bones that, for a variety of documented and undocumented reasons, are missing from his phantom torso, leg, and arm.[87] Even the outstretched hand that initially drew me to Stuart Gunn is not entirely his. Years later, I learned that because he had "suffered significant weight loss prior to death," researchers did not consider his body to be "of a normal size."[88] Thus, that boxed arm—those reaching fingers—were cast from a "healthy . . . living substitute," erasing illness from Stu's phantom form.[89]

Again, it is this tension—the exposed body as both constructed and more than construct—that haunts nuclear safety. The phantom is at once an abstract model and concrete calibration device, translating life into risk and risk into a broader politics of accountability. It links subject and structure within a powerfully coded transpositional frame, administering protection in probability.

Such renderings are not absolute, but co-constituted by representational fissures and gaps. As philosopher Judith Butler argues, "to be a body is to be exposed to social crafting and form" that influences one's very legibility.[90] "The epistemological capacity to apprehend a life is partially dependent on that life being produced according to norms that qualify it as a life, or, indeed, as part of life," they write. These norms make some people "recognizable" and others "decidedly more difficult to recognize."[91]

Re-membering the exposed body, therefore, means reckoning with both the human product and its hauntings. It means expanding "the domain of the empirical" to include the conditions of memory itself—not just what (who) is remembered but how, in what form, and to what end.[92] Such ghost stories seek "not only to repair representational mistakes," Gordon writes, but to create a more equitable "countermemory, for the future."[93] They evoke the living-being that exceeds its normative possibilities, unsettling historical divisions between the rational and imaginary.

3 Rational Mutants

Imagine: sixteen preeminent scientists sitting around a conference table calculating the mutation of the human race. They sip coffee and smoke cigarettes and page through government reports about nuclear fallout. They make notations and scratch figures in the margins. One complains that the AEC hasn't provided them with adequate data and wonders derisively what the agency is hiding. Another sighs in exasperation at the comment. Choice words are muttered under breaths. It is February 1956. A cold, dry winter in Chicago. The bomb is ten years old.

The scientists are frustrated: by the impossibility of their task, by the weighty importance of it nonetheless, by the fact that they are serving on a high-profile committee with professional rivals. The meeting has been contentious from the start, crowded with egos and resentments.[1] Some are vocal proponents of aboveground nuclear testing, while others are wary of the practice. A few have publicly criticized the AEC and its policies about nuclear safety. A few have worked for the AEC and authored those policies.

However, the scientists concur on one essential point: radiation causes genetic harm, and therefore the atomic age may be changing humanity. The scale and consequences of that change are up for debate, they argue, but the problem deserves careful attention. This shared conviction keeps them at the table with their cigarettes and coffee, tracing mutant futures. It keeps them talking to one another, albeit sometimes through gritted teeth.

I have spent years imagining that committee meeting. Countless hours at my desk reading and rereading the transcript of the scientists' discussion, and still more studying their recommendations for radiogenic exposure. It would be fair to say that, in the process of writing this book, those two February days became a minor obsession for me.

It was a sentence on the second to last page of the committee's final report that kept drawing me back. Actually, it was a phrase within a sentence—a parenthetical hedge—that I just couldn't seem to shake. There, in the conclusion, the committee recommended that individuals receive no more than ten roentgens of man-made radiation before the age of thirty.[2] This dose, they argued, was "reasonable (not <u>harmless</u>, mind you, but <u>reasonable</u>)"[3] because, while millions would become mutant as a result of their exposures, the human race as a whole would survive.[4]

The scientists discussed the ten roentgens figure at length, but I was less interested in the number than in the caveat that accompanied it. Instead, it was those six words (*not <u>harmless</u>, mind you, but <u>reasonable</u>*) that continued to hold my attention. It was the notion of rational mutation itself, the explicit coupling of reason and harm, that brought me back to the table again and again.[5]

I first encountered the committee's final report in the summer of 2013, while making my way through the Ava Helen and Linus Pauling papers at Oregon State University. The Paulings were outspoken about the dangers of nuclear fallout during the 1950s and 1960s, and reading their archive was like stepping into the middle of a heated debate. There was the full-page ad in the *New York Times* shouting, "Is this what it's coming to?" beneath a milk bottle marked with a skull and crossbones.[6] There was the op-ed in the *Saturday Review* calling for an end to aboveground testing that warned, "There is no known way of washing the sky."[7] There was the statement from *Playboy*'s editorial board condemning "the contaminators" who they argued were threatening life on Earth. "Alarmist talk?" they asked. "Yes. It is time for alarm."[8]

Much of the archive focused on environmental monitoring: academic articles that traced the bomb on wind and water currents, government reports that measured it in thundershowers and human bones. I was surprised to learn that Hanford scientists had spent two summers in the high arctic running entire villages of people through whole-body radiation counters.[9] And that a team from UCLA had tramped through the Amazon rainforest recording atomic signatures in three-toed sloths, kinkajous, and wild dogs.[10]

There were studies about fallout on rooftops and city streets, fallout in wheat fields and grocery stores, fallout in bottles of beer and puddles of rain. Scientists found it in Cuban tobacco and Arctic polar bears and Chesapeake Bay oysters and Wisconsin cheese platters and New Mexican hamburgers and thousands of baby teeth donated by St. Louis area children. They found it in herds of sheep just outside of Damascus and handfuls of dandelions in upstate New York and families of reindeer wandering the frigid Lapland plains. And they found it in dust. Everywhere, in dust. Nigerian dust. Australian dust. South African dust. Brazilian dust. German dust. Canadian dust. Chilean dust. The list went on and on.[11]

The word *fallout* is a product of the atomic age, a term that has become synonymous with aftermath.[12] But those articles and reports demonstrated the embodied ways that fallout exceeded its definitions. There was no after to the bomb in that archive. The mushroom cloud hadn't disappeared; it had only dispersed. Families drank the bomb with each glass of milk. When a child tied a string to a loose tooth and gave it a yank, the bomb popped out of their mouth.

At first, this ubiquity was a revelation within the scientific community. "Nobody believed you could contaminate the world from one spot," recalled Lester Van Middlesworth, who spent part of his career measuring fallout in farm animals. "It was like Columbus when no one believed the world was round."[13]

Perhaps the most formative event during this period was a thermonuclear detonation called Castle-Bravo in the spring of 1954.[14] The largest U.S. test to date, Bravo had the explosive power of one thousand Hiroshima bombs and covered the nearby Marshall Islands with so much ash it looked like a blanket of fresh snow.[15] The ash also fell on a Japanese fishing boat, the *Daigo Fukuryu Maru* (*Lucky Dragon No. 5*), which had been catching tuna eighty-five miles away. When the crew returned to port two weeks later—gums bleeding, eyes oozing, and hair falling out—they quickly made international news. The Japanese government responded by initiating an inspection protocol for all incoming fishing boats and sponsoring a large-scale study of Pacific fisheries. Over the next two months, study scientists identified fish with elevated radioactivity levels thousands of kilometers away.[16]

Bravo raised new public concerns about nuclear weapons and their effects. While it was common knowledge that some Hiroshima and Nagasaki survivors suffered from *atomic bomb disease* (symptoms from high-level radiation generated during an explosion), Japanese medical investigators reported that the *Lucky Dragon*'s crew were experiencing something else: a syndrome they called *acute radiation disease*. The critical difference, they explained, was that the damage did not result from the initial blast, but from the debris that it had produced.[17] It was a hazard that could be borne on the wind for thousands of miles, an illness that fell from the sky like rain. Japanese media outlets dubbed it *shi no hai* (ashes of death), and around the world, *fallout* became a household word.[18]

The AEC strongly resisted such language, arguing that Bravo's byproducts were not dangerous. Chairman Lewis Strauss stated that while nuclear testing had indeed raised background radiation to some degree, it remained "far below the levels which could be harmful in any way to human beings."[19] His words rang hollow against the news reports continuing to emerge from Japan and the Marshall Islands. When a member of the *Lucky Dragon*'s crew died soon after the event, one Japanese newspaper wrote, "Has the death of a citizen ever been watched by so many eyes? They are the eyes of a strong anger and protest against the ashes of death."[20] That same month, Godzilla emerged from the sea and onto the silver screen, his mutant skin dripping with radioactive fury.[21]

By the end of 1954 the Rockefeller Foundation had entered the fallout debate, arguing that the public needed an independent scientific evaluation they could trust. It selected the National Academy of Sciences (NAS) to organize the assessment, funding an extensive year-long study entitled the Biological Effects of Atomic Radiation (later known as the BEAR report).[22] In a press release, the foundation called it "a dispassionate and objective effort to clarify the issues, which are of grave concern and great hope to mankind."[23]

Initially, NAS president Detlev Bronk was reluctant to take on a project of such magnitude. When a Rockefeller trustee suggested over cocktails one evening that Bronk should "do something" about fallout, he replied, "That is a very broad request, because what do you do about it?"[24] I can't help but smile at his candor in that moment. When I picture it, I place Bronk and the trustee in a Manhattan boardroom and hand them each a heavy glass with brown liquor and ice. Sometimes I imagine Bronk's response is immediate and terse, a brusqueness borne of irritation and overwork. At other times I allow him to pause before answering, letting the question hang awkwardly between them in the air. In my favorite version, however, the suggestion that he *do something* about fallout makes Bronk choke on his drink. When he finally speaks, he is all incredulous laughter and sputtering.

Because, really, what *do* you do about fallout? Where do you begin, when your subject is both invisible and ubiquitous? How do you "clarify the issues" for an anxious public (dispassionately and objectively), when the problem is already humming inside of their bones?

The bomb-as-fallout presented a novel set of spatial and temporal challenges. Instead of annihilating thousands of people in a single, explosive moment, it picked them off slowly through stillbirths and leukemia, thyroid disorders and intergenerational risk.[25] This attritional form of violence was disorienting, making it difficult to connect fallout with its effects. As one member of a Chernobyl documentary crew said years after the disaster, "It wasn't obvious what to film. Nothing was blowing up anywhere."[26]

Ultimately, however, Bronk agreed to coordinate a scientific reckoning of global proportions. Throughout 1955 and 1956 he convened committees in six key areas: genetics, pathology, meteorology, oceanography and fisheries, agriculture and food supplies, and the disposal and dispersal of radioactive wastes. He populated each with scientists from both academia and government, several of whom worked for the AEC or had received research funding from the agency. Though officially tasked with assessing radiation's biological effects, the committees also articulated a broader argument about the "reason" of exposure itself.

This was particularly true for the genetics committee, which found little to debate when it came to radiation's mutagenic properties.[27] The question wasn't *whether* radiation produced genetic harm, its members agreed, but how to parse the relationship between injury and progress. Impeding nuclear development "might seriously weaken our country's position in the world," they argued, or it "might deny future generations some of the possible benefits of nuclear power and other atomic discoveries."[28] Thus, rather than avoiding exposure altogether, they sought to adequately "balance risk against risk."[29]

Just how to achieve such balance in practice, however, was far more ambiguous. Though the committee recommended ten roentgens as a reasonable exposure limit, it also made the case that reason was situational. A nation at war, for example, might be willing to accept higher levels of exposure in exchange for the promise of greater national security. And increased mutation rates might be the price that society had to pay for the potential rewards of nuclear industry.

However, even as the committee described nuclear development as "desirable or even almost obligatory," it emphasized its significant costs:[30]

> The basic fact is—and no competent persons doubt this—that radiations produce mutations and that mutations are in general harmful. . . . Different geneticists prefer differing ways of describing this situation: But they all come out with the unanimous conclusion that the potential danger is great. . . . We ought to keep all of our expenditures of radiation as low as possible. . . . From the point of view of genetics, they are all bad.[31]

The committee argued, therefore, that investing in nuclear technology meant existing in contradiction. It warned against radiogenic exposure while simultaneously arguing that this was integral to the nation's development, framing public health and economic well-being in opposition and asking Americans to accept "sensible" trade-offs between the two.[32] In the process, it lent scientific authority to the Cold War notion that death could be rational—*or even almost obligatory*—when it was for the greater good. Not harmless, mind you, but reasonable.

LOST IN THE CROWD

Though the genetics committee presented a unified message to the public in its final report, meeting transcripts and personal correspondence told a different story. It took months of negotiation—first in person, then via telephone and post—for members to reach an eventual, grudging consensus. Two of the committee's most prominent geneticists (Herman Muller and Sewall Wright), for example, were barely on speaking terms throughout the deliberation process, each threatening to remove his name from the final report just weeks before publication.[33] Other members disagreed whether the committee should provide reasonable exposure limits at all, given that no exposure was recommended from a genetics perspective.

"I for one am unprepared to sign any report that states a permissible dose," geneticist James Neel argued at the Chicago meeting, "I feel we should not compromise ourselves as scientists."

"I think we all inwardly rebelled against the idea of having to give an answer, as scientists, to a question that can't be answered scientifically," geneticist William Russell responded. "However, for practical purposes someone has to set this dose now. . . . [I]f we don't do it, somebody else is going to do it."[34]

Committee chairman and mathematician Warren Weaver insisted that "practical considerations must probably dominate" their discussion.

"We have to face the question: 'What are proper safeguards?'" he advised. "It won't be sufficient to say: 'As little [exposure] as possible.' That kind of statement won't be helpful."

"The most important thing is to set a value, conservatively but reasonably, so it will not have to be discarded [by the AEC] as impractical," physicist Gioacchino Failla recommended.

"Then what we need to know is what is practical," geneticist Alfred Sturtevant replied, "so we can decide what we can reasonably suggest."[35]

However, practicality, like reason, proved a slippery target. Committee members disagreed about what constituted a practical exposure limit and how much weight to give industry concerns about cost.

"What if they [AEC officials] say it is going to be awfully inconvenient? Will we shade our judgment? Do we doubt what we want?" geneticist C. C. Little asked.

"We are necessarily forced into the position of weighing the possible benefits, that is, the risk from not doing things, such as national defense and atomic energy, against the genetic factors, and somebody has to make this kind of decision," Russell answered. "Maybe we should assume the responsibility, and if we do, then we must take these things into consideration."

"You can't separate the two things realistically," Weaver said. "You say: 'It would be pretty costly [economically] to go down this far, but....' At that point you look over to the other argument, and you say: 'What kind of a genetic price would we pay if we didn't go that far?'"[36]

Viewing exposure through the lens of practicality meant asking not only how many lives *could* be lost to atomic development, but how many lives *should* be lost to it. Several committee members bristled at this framing. What would it mean to recommend a specific amount of injury and even death? And how, as scientists, could they justify such a claim? "I don't think we can produce a magical figure and throw it to the public," Weaver said. "Now how far we go, and just what kind of justification it is reasonable to give to the general public, is a very complicated and difficult question."[37]

Geneticist George Beadle worried that readers would misunderstand what the committee meant by reasonable exposure. In part, his concern had to do with the industry term *permissible dose*, which he found both opaque and deceptive. Formalized in the *National Bureau of Standards Handbook* soon after the Bravo test, permissible dose framed bodily harm as a probabilistic state that could be weighed against other assessment parameters.[38] As the NCRP put it, "The concept of a permissible dose envisages the *possibility* of radiation injury manifestable during the lifetime of the exposed individual or in subsequent generations. However, the *probability* of the occurrence of such injuries must be so low that the risk would be readily acceptable to the average individual" (italics in the original).[39]

While he understood the NCRP's rationale, Beadle found this new terminology problematic. "A beautiful example of a true, but misleading, statement comes right out of [AEC commissioner Willard] Libby's remarks. He says the amount of radioactive fallout is only a small fraction of the permissible dose. That is accurate, but completely misleading, because he doesn't say that 'permissible dose' doesn't mean a thing at all."[40]

Though committee members agreed that permissibility was relative, they disagreed about how they should account for such relativity in their own recommendations. Averaging exposure across an entire population, for example, would mask its uneven distribution and make individual risk appear less significant. So too, the amount of time they considered—one generation of people versus ten—would necessarily influence the scale of impact they found.

"Is the really serious matter the heavy radiation of a few individuals or is it rather the exposure of everybody to small amounts of radiation from fallout?" Weaver asked. "Is the problem just one of damage to individuals, or is there a danger point for the population?"

"From the population's standpoint, certainly in this country with its high mobility of population, the effect would simply dissolve in the mass . . ." Wright replied. ". . . the dose received by the individual really doesn't mean anything. It's the total dose received by the population that matters."

"Yes," Muller interjected sarcastically, "but we felt it was a little unfair to the person and his descendants to pile it up on him, unless he volunteers."

"I would like to ask," Weaver said to Wright, "whether one who has worked in population genetics for a long time isn't likely to forget that the units of population are, after all, individual people. I don't think you capture all the problem in pointing out to a person who suffers from an induced mutation that he is only one of 350,000,000 people. He says, 'Yes, but I am me. I am still me, and I have interests, and I have responsibilities.'"[41]

Still, Weaver maintained that the committee should give the public "some sense" of impact on a national scale. In that spirit, he asked each member to take time on their own after the meeting to estimate the level of damage associated with a ten roentgen exposure limit.

"Now here is a bunch of babies that the public understands about and will be concerned about," he said, "not just x indefinite babies 1000 years from now. Assume that all the people now alive [in the United States] receive a certain dose of radiation, say 10 r units. Let's try to estimate what that would do to this bunch of 160,000,000 babies. Secondly, let's estimate what it will do to their descendants up to, but not beyond, the tenth generation, that is F_2 through F_{10}. We would come up with an estimate that the total dose of 10 r administered to the population of 160,000,000 would produce _____ (lower estimate) to _____ (upper estimate) detrimentally affected persons."[42]

For the most part, committee members agreed to the temporal boundaries that Weaver suggested. Though they recognized that the bomb's mutagenic effects would extend beyond ten generations, the uncertainties associated with deep time would eventually become too great.

"We cannot foresee what kinds of social changes may be coming which would so terrifically alter the estimation for more remote generations as to make the estimates meaningless," geneticist Tracy Sonneborn said in support.

"Yes, if we began to talk about what would require 20 generations, or 40 or 100 generations, we would find ourselves essentially in the position—the embarrassing position—of playing a long-range god in this matter," Weaver agreed. "It's difficult enough to play short-range god in this business, but to play a long-range god is impossible."[43]

When it came to defining *detrimentally affected persons*, however, committee members again found themselves at odds. Wright did not think they should include miscarriages in their assessment, focusing their attention on children with developmental disabilities instead. "It seems to me," he argued, "that what we want to know is the risk run by the person who gets a dose, of having a child that is, say, feeble-minded, or has muscular dystrophy, or some other distressing danger. We don't care about lethals. The prenatal lethals nobody will ever know anything about, and they're unimportant."[44]

In addition, Russell and Muller wondered how to account for exposure's less visible effects, such as slight reductions in life span. A loss of several months or even years would likely go unnoticed at an individual level, but did that mean that the committee should disregard it? Where should they draw the line between a significant impact and an insignificant one?

"Would you include mutations that reduce the life span by 5 per cent and which would not appear visibly?" Russell asked. "Would you estimate this in the calculation and tell a person he was taking that risk?"

"What about a mutation rate that gives, say a 5 per cent risk of death before reproduction, but if it does not succeed in doing that, has no significant effect?" Muller added.

"I would say that is a definite handicap," Weaver replied.

"We need to bear in mind that for a given mutation we can't set a limit as to what is significant, and have a threshold," Muller continued. "Even though things are not additive, they are cumulative in some sense. Like dust—one grain of dust won't bother you, but a dust bowl will."[45]

Of the committee's sixteen members, only six ended up making the calculations that Weaver requested. Wright, Neel, and geneticist Milislav

Demerec refused to participate at all, arguing that the task "relied on too many unknown quantities" and would not provide a useful estimate of genetic impact.[46] And indeed, the six members who completed the exercise produced such a wide range of figures that geneticist James Crow did not think they should be included in the committee's final report. Instead, he made the case that a "best estimate or some narrow range of estimates" would have greater explanatory power.[47]

In the end, the committee settled on five million "mutants" as the "most probable estimate."[48] However, it clarified in its final report that this impact would only be visible "in a statistical sense."[49] Not only would such damage be distributed throughout time and space, it would be impossible to identify individually. The "children born to exposed parents . . . who would be definitely handicapped because of the mutant genes due to radiation" would be indistinguishable from those with disabilities unrelated to atomic development.[50] While statistical mutations could be counted, in other words, the people living and dying with them could not. The effects of permissible dose would be, "so to speak, lost in the crowd."[51]

REGULATION

When I read the committee's final report in the Paulings' collection that summer, those words—*lost in the crowd*—rang through me like a bell. The idea that millions would be "lost" felt jarring in an archive so thoroughly invested in visualizing the invisible. Because of fallout, scientists could follow tiny specks of dust as they traveled from one side of the globe to the other. This allowed them to map air currents and ocean gyres and predator-prey relationships in detail, transforming the very nature of what could be seen.[52]

And it was the intimacy of those visualizations—the banal, everyday descriptions of contaminated life—that I had found so compelling about the archive in the first place. It wasn't just that fallout was everywhere, but that everywhere happened in places. In homes. In lungs. In baby teeth. In brows furrowed with curiosity and concern. Within those pages, fallout was more than an abstract by-product of the bomb. It was a lived condition, uneven and specific.

However, while this lived specificity was a boon to Cold War scientists, it proved challenging to regulators. Though the federal government had banned "injurious" chemical additives in food during the first half of the twentieth century, it had yet to address environmental contamination more broadly.[53] So too, the definition of safety that informed food production and occupational health in the 1950s did not make sense for nuclear materials. While industrial chemicals were considered "safe" below a particular threshold at that time, there was no genetically harmless amount of radiation.[54] Instead, injury was inherent to the technology itself, requiring an altogether different regulatory calculus.

In addition, risk-based policies framed exposure as a matter of choice. Permissible dose, after all, did not promise absolute protection; it sanctioned probabilistic impact at levels "readily acceptable to the average individual."[55] This effectively positioned everyday Americans as the arbiters of reasonable harm, making risk management a personal responsibility.

However, though it emphasized risk thinking in the abstract, permissible dose provided little practical advice for the American public. When, for example, was it rational to refuse or "readily accept" contamination in one's food and children? What would that actually look like in daily practice? For that matter, what did it mean to reckon with the mere *probability* of disease?

Among the academic studies, government reports, and op-eds in the Paulings' collection, I found thick stacks of letters asking about this very thing. Some were from mothers who wanted to know how to feed their families in a reasonably contaminated world. Others were from doctors wondering how to advise patients about avoiding "overexposure."[56] A grandmother from Michigan wrote Linus Pauling to inquire whether fallout could have caused her granddaughter's leukemia.[57] A woman from Los Angeles asked if members of her church might "check" themselves for radiation using Geiger counters or something similar. "Those who have received dangerous amounts can then be more realistically aware of their situation," she hoped. "Others who have not accumulated serious amounts will have their fears allayed."[58]

Like the fallout studies that inspired them, these letters described the urgent yet ordinary details of contaminated life. Together, they narrated the emotional labor of acceptable risk: a distinct combination of hypervigilance and self-doubt.

Dear Dr. Pawling [sic],
The question I wish to ask you may be silly only in the actual statistical difference, which may be so infinitesimal as not to make any important difference. It is... I am breastfeeding my child, and wish to do so for another two months. The child is now 2½ months old. Is there more Strontium 90 in my milk or the cows? And if there is less in mine would it be healthier to nurse the baby a little longer?[59]

Dear Dr. Pauling, ...
Perhaps I should tell you how our family attempts to take precautions....

- The purchase, last Sept., of 150 pounds of locally produced powdered milk. I was assured the milk had been processed before the Soviet tests resumed. No fresh milk purchased since that time, although we are not so rigid we do not allow children to drink fresh milk in other homes, from time to time....
- Sticking to well-aged cheese (the adults in our household prefer it anyway). I have been talking about freezing cottage cheese for a month or longer....
- Abandoning the use of wheat germ (which I had erratically tossed into meat loaf, etc. from time to time).
- Not making a game of playing in the rain. Discouraging the drinking of rain water. (When we were able to find out what the background radiation was, we tended to keep the children indoors—in our wooden house—where feasible, when the levels were over .03 milliroentgens per hour)....

Could you comment on these procedures?[60]

Dear Dr. Pauling,
I feel presumptuous imposing my insignificant request upon a man so busy.... Many of my friends think I am foolish to be concerned about this whole thing, however, if someone were holding a gun on my children, I would fight, and when one comes right down to it, what is the difference?[61]

In addition to their detailed descriptions, I was struck by the writers' self-consciousness—the questions beneath the questions that tugged at their letters like an undertow. *Is it safe to nurse my child? (And is this a silly thing to ask)? Will this meal hurt my family? (My friends think I'm being foolish; do you)?* The more I read, the more I noticed the doubled nature of their requests—how milk and meat loaf became proxies for emotional control. Because if acceptable risk was defined by one's willingness to accept it, then living with fallout required more than managing one's exposure. It also meant embodying the logics of rational mutation: a physical and existential remaking of the self.[62]

In 1955 the Federal Civil Defense Administration (FCDA) distributed an instructional pamphlet called *Facts about Fallout* to communities across the United States.[63] It featured a nervous cartoon man beneath a mushroom-clouded sky who had just learned that he, like every American, was "a potential target for fallout." The risk may be small, the pamphlet explained, or it could be catastrophic. "If you are exposed to it long enough—IT WILL HURT YOU! IT MAY EVEN KILL YOU!"

Alarmed, the man searched in vain for the source of the danger. First, he listened for fallout, then scanned the horizon for it with his binoculars. Next, he held a butterfly net, hopefully, to the sky. "You can't hear it," the pamphlet warned, "You can't taste it. You can't touch it. You can't smell it. Often you can't even see it." Disheartened, the man slumped against a wall and hugged his knees to his chest. "Don't get discouraged," the pamphlet admonished. He stood up and ran back and forth, uncertain what to do. "Don't get panicky," it told him.

The best protection was to be prepared: to stock the basement "with food staples as Grandmother did." Or better yet, to build a fallout shelter with beds and games for the children, a place the entire family could live comfortably until the danger had passed. With relief, the man retired to his now well-stocked basement, where he relaxed with a book, his expression calm for the first time. "Americans are hard to scare," the pamphlet concluded. "Of course, we are worried about the forces science has unlocked. We would not be intelligent human beings otherwise. But this problem can be solved—as others have been—by American ingenuity and careful preparation. Now, see your local Civil Defense office.... Let Civil Defense help you to help yourself."

Facts about Fallout was part of a broader FCDA campaign to "emotionally manage" the American public.[64] Its goal was not to irradicate fear altogether but to calibrate it for life in the nuclear age. Narratives of atomic threat "had to be formidable enough to affectively mobilize citizens," Joseph Masco writes, "but not so terrifying as to invalidate the concept of defense altogether." Instead, the FCDA sought a "contradictory state of productive fear," one that normalized the ever-present possibility of catastrophe.[65] As FCDA director Val Peterson told *Collier's* in 1953, "Fear can be healthy if you know how to use it."[66]

The aim of civil defense, therefore, was to maintain "healthy" levels of fear without inciting mass panic. FCDA films like *Let's Face It, Bombproof,* and *Target You* emphasized the strategic utility of self-control, framing it as the last line of American defense.[67] "Like the A-bomb," Peterson said, "panic is fissionable. It can produce a chain reaction more deeply destructive than any explosive known. If there is an ultimate weapon, it may be mass panic—not the A-bomb . . . every citizen [is] a target."[68] Emotional hygiene, he continued, was essential to national security—the "preventative medicine" that would ensure the nation's survival. The most effective protection against the bomb could be found within: *Let Civil Defense help you to help yourself.*[69]

Critically, the self that civil defense imagined was one that could navigate the probabilities of nuclear exposure. For although fallout shelters offered a significant degree of protection, the FCDA warned, Americans could still "soak up a serious dose" of radioactivity in the aftermath of a bombing.[70] If that happened, it explained in its 1951 pamphlet *Survival under Atomic Attack,*

> Then you most likely would get sick at your stomach and begin to vomit. . . . For a few days, you might continue to feel below par and about 2 weeks later most of your hair might fall out. By the time you lost your hair you would be good and sick. But in spite of it all, you would still stand a better than even chance of making a complete recovery, including having your hair grow in again.[71]

The FCDA described radiation sickness as a collection of embodied symptoms and "calculated chances."[72] Though a *serious dose* could make a person *good and sick*, they would still have *a better than even chance* of *complete recovery*. Greater than 50 percent. More likely than not. Better odds than a flip of a coin.

Such statistical reckoning, the FCDA argued, was essential to "bombproofing" the American public against panic. It tempered the chaos of nuclear war through quantification, presenting survival as the measurable product of likelihood and luck.

> Should you happen to be one of the unlucky people right under the bomb, there is practically no hope of living through it. In fact, anywhere within one-half mile of the center of the explosion, your chances of escaping are about 1 out of 10. On the other hand, and this is the important point, from

one-half mile to 1 mile away, you have a 50-50 chance. From 1 to 1½ miles out, the odds that you will be killed are only 15 in 100. And at points from 1½ to 2 miles away, deaths drop down to only 2 or 3 out of each 100.[73]

However, the FCDA also acknowledged that survivors weren't likely to escape unharmed. "Naturally, your chances of being injured are far greater than your chances of being killed," it clarified. "But even injury by radioactivity does not mean you will be left a cripple, or doomed to die an early death. Your chances of making a complete recovery are much the same as for everyday accidents."[74] And although the longer-term effects of an explosion like "lingering radioactivity . . . may be dangerous," it continued, "still it is no more to be feared than typhoid fever or other diseases that sometimes follow major disasters. The only difference is that we can't now ward it off with a shot in the arm; you must simply take the known steps to avoid it."[75]

In addition to building fallout shelters, this included good atomic housekeeping like closing windows and doors for "at least several hours" after a bombing and laundering contaminated clothing with "warm water and plenty of soap."[76] These "simple steps," it assured readers, would "go a long way toward keeping your house from being contaminated by lingering radioactive wastes scattered about."[77] Still, no amount of cleaning would be completely effective. "Whenever there is widespread neighborhood pollution, it will be impossible to keep your house absolutely free of it. A little is bound to seep in through cracks."[78] Above all, the most important thing was to keep calm. "Take these precautions, but don't worry. There isn't much chance really dangerous amounts will pile up in the house."[79]

Like most civil defense materials, *Survival under Atomic Attack* did not clarify what it meant by "really dangerous amounts," nor did it address radiation's mutagenic properties. Instead, it echoed the federal government's call for "emotional inoculation" and "psychological defense" against nuclear risk.[80] "If you follow the pointers in this little booklet," it concluded, "you stand a far better than even chance of surviving the bomb's blast, heat, and radioactivity. . . . But if you lose your head and blindly attempt to run from the dangers, you may touch off a panic that will cost your life."[81] Just as atomic weapons represented both problem and solution in Cold War geopolitics, civil defense framed the self as the nation's greatest threat *and* the key to its salvation. The difference, it seemed, was one of resolution—a matter of turning the existential dial just right.

COMMON SENSE

When the NAS published the BEAR report in June 1956, it did not "clarify the issues" as the Rockefeller Foundation had hoped.[82] Rather, it joined the tangle of mixed messages that Americans were receiving from scientists, civil defense officials, media outlets, and the AEC. In part, the report contributed to this confusion by presenting what seemed like contradictory arguments. While the pathology committee downplayed the risks of low-level exposures, the genetics committee highlighted them, refuting federal assurances that fallout from nuclear testing was safe. Instead, it argued that when it came to atomic development, the very definition of safety had changed.

This came as a surprise for many Americans, especially those schooled in the language and logic of civil defense. As Edward P. Morgan said in his nightly ABC newscast, "There had been a misconception that there was a socalled [*sic*] 'safe' level in exposure to radiation, below which effects would be harmless. Now we have it on the authority of some of the most outstanding scientists in the land that no matter how tiny a dose you get, radiation not only can harm you but all your descendants as well."[83] Reporter Dorothy Thompson echoed this sentiment in *Ladies Home Journal* several months later, calling the report "one of the most important events of the year."[84] The genetics committee had unsettled the very idea of radiation protection, she wrote. When "asking themselves the question: 'What is overexposure?' they answer, 'All exposure that can conceivably be avoided.'" Any amount of radiation was too much genetically, she concluded. It was "harmful in the most enduring and incurable way."[85]

However, though the genetics committee challenged the AEC's claims about safe exposure, it supported a broader regulatory shift toward risk-based protection. The costs and benefits of atomic development had to be weighed rationally, it told readers. Exposure could only "conceivably be avoided" if such measures did not endanger national security and economic development.

The BEAR report made clear, therefore, that nuclear safety limits were inherently social—a fact that many readers found shocking. "It may perhaps sound startling to contend that the establishment of permissible levels of radiation exposure is not basically a scientific problem," NCRP president Lauriston Taylor said in response to the report. "Indeed, it is more a matter of philosophy, of morality, and of sheer wisdom.... [P]resent and future solutions of the radiation protection problem will have to be based on a risk philosophy; they will have to be compromise solutions, and cannot be solved on the basis of scientific evidence alone."[86]

Like many in the field of radiation protection, Taylor saw increased regulation as a potential threat to nuclear production. Strict safety standards "could be very costly and could seriously retard the atomic industry," he told Congress's Joint Committee on Atomic Energy (JCAE) in 1957.[87] "There is a very real danger, if one keeps going down and down [i.e., lowering permissible exposure limits] that you will price us out of the adequate use of this new and very valuable tool."[88] The federal government must maintain a "common sense" approach to radiation safety, he insisted, or risk being "economically purified out of business."[89]

So too, Taylor believed that nuclear workers (who had higher permissible dose limits than the general public) "should expect to take their share of the risk involved in such a philosophy."[90] As he explained in a letter to AEC chairman Strauss,

> I see no alternative but to assume that [an] operation is safe until it is proven to be unsafe. It is recognized that in order to demonstrate an unsafe condition you may have to sacrifice someone. This does not seem fair on the one hand, and yet I see no alternative. You certainly cannot penalize research and industry on the suspicion of someone who doesn't know by assuming that all installations are unsafe until proven safe.... [Nuclear development] provides work for the worker and without this the worker might be without a job. If the worker demands an absolute guarantee of safety or benefits for imagined and unproven dangers, he penalizes the industry to the point where industry cannot operate.[91]

For Taylor and most radiation professionals at the time, a commonsense approach was necessarily a comparative one.[92] Protection, they advised, should balance exposure-based risks against those the "average person" might encounter in their "ordinary pursuits."[93] AEC commissioner

Willard Libby reiterated this philosophy before the JCAE, equating ionizing radiation with other everyday hazards:

> It is not contended that there is no risk. But all life, and every minute of our day and night, is measured in terms of risk. . . .We make our choice: How much risk are we willing to take as payment for our pleasures—swimming at the seashore, for example—our comfort, or our material progress? Here our choice seems much clearer. Are we willing to take this very small and rigidly controlled risk, or would we prefer to run the risk of annihilation which might result if we surrendered the weapons which are so essential to our freedom and our actual survival?[94]

Though Taylor, Libby, and their contemporaries acknowledged that exposure limits were social, they did not go so far as to address their structural inequalities. On the contrary, safety standards based on averages explicitly ignored the bomb's uneven effects. "These things are to be understood quantitatively, with appreciation of the fact that the harm done by small amounts of radiation is out of reach of direct observation," radiologist R. R. Newell wrote in response to the BEAR report. Mutagenic injuries would "forever be untraceable" to the source, measured and managed only in abstraction. The challenge was to think about the problem "logically, not emotionally," he advised. As long as average exposures remained low, individuals needn't worry "about the occasional bad chance."[95]

Thus, nuclear safety valued individual people as rational choosers even as it devalued them as statistically insignificant. Such paradoxes, AEC Health and Safety Laboratory director Merril Eisenbud explained, were integral to the very philosophy of permissible dose. "One person says that the risk is negligible: what he means is that the risk being one in a million, or one in ten million, it is so small that the individual is in no great danger," he wrote in 1961. "Another person insists that the risk is serious, for he notes that a risk of one in a million, applied to a population of three billion, will produce many tragedies. Both are correct, from different points of view."[96]

Industry professionals made strategic use of the contradiction that atomic development could produce *many tragedies* while presenting *no great danger*. Tacking back and forth between human and statistic, choice and chance, they framed exposure as an abstract, affective condition.

Protection meant accepting risks that would only ever (officially, legibly) injure statistical people. It meant monitoring one's dose as well as one's reaction to its probabilistic effects, asserting emotional control over the quantitative self. And it meant normalizing the tautology of reasonable harm: that an acceptable risk was an accepted one.

4 Body Burden

On February 3, 2000, more than five hundred Hanford workers and their families crowded into Richland's Federal Building auditorium. They found seats, or stood when there were no seats left, offering chairs to the elderly and infirm. Then they walked to the front of the room, one by one, and told stories about exposure and illness.

First to the microphone was a retired worker with gray-white hair, glasses, and lung cancer. Next, a daughter spoke for her father, who died of bone marrow disease at age fifty-nine. Third, a middle-aged man wearing anger and a plaid shirt described the tumors that grew on his hand, lung, and adrenal gland. He held up a form he was required to sign allowing him to exceed his lifetime dose of radiation. Four: pancreatic, lung, and adrenal cancer. Five: prostate cancer and asbestosis. Six: lymphocytic leukemia. Seven: asbestosis and beryllium disease.

The succession of stories continued for hours as hundreds of bodies stepped onto the stage. *"I've been diagnosed with cancer of the pancreas with six weeks to live, God save me."* Hundreds of workers licked lips and cleared throats that caught with nerves and emotion. *"They say the only thing they can do for me is a lung transplant. Any of you got a lung you want to get rid of?"* Hundreds of hands unfolded prepared statements

A mushroom cloud at Richland High celebrates the school's mascot, 2016. Photo by author.

to be read aloud. "*I did my work and I did it proudly, what will you do for us?*" Hundreds of people, indelibly marked by the bomb, took a deep breath and spoke.[1]

The evening was historic. For many, it was the first time they had shared their concerns publicly, an uncommon act in this pronuclear place.[2] "No one ever thought they'd see it happen," one reporter wrote of the event. "Not in this town."[3] This town that coaxed plutonium from uranium slugs for more than four decades to power the nation's nuclear arsenal. This town that cheered for a high school football team called the Bombers whose helmets were decorated with mushroom clouds.[4] This town that now lived and worked alongside the nation's largest nuclear dump.[5]

Richland was one of nine stops on what DOE had dubbed its "listening tour" of the weapons complex. Organized in response to a "massive and unprecedented review of worker health studies at nuclear weapons plants," the agency had recently acknowledged that there was "credible

A bumper sticker in Richland expresses pride about the city's atomic history, 2012. Photo by author.

evidence" its employees "may be at increased risk of illness from occupational exposures to ionizing radiation and other chemical and physical hazards."[6] The local *Tri-City Herald* wrote that the news "hit Hanford like a lightning bolt from one of the sudden summer storms that sweep over the nuclear reservation."[7] Perhaps a more fitting analogy would be that the news hit Hanford like a bomb.

In the months that followed, the Clinton administration promised a new era of transparency and accountability. As DOE secretary Bill Richardson told the *Washington Post*, "The Department of Energy is coming clean with its workers."[8] On April 13, 2000, Vice President Al Gore pledged federal support for the sick and injured, saying, "Today this administration begins the process of healing by admitting the government's mistakes, designing a process for compensating these workers for their suffering, and by becoming an advocate for Department of Energy workers throughout the nuclear weapons complex."[9] Six months later, Congress passed the Energy Employees Occupational Illness Compensation Program Act (EEOICPA), widely considered a landmark piece of legislation.[10]

EEOICPA expanded the legal terrain of the nuclear battlefield. Though DOE workers produced the technologies of war, they had lacked the veteran status, medical benefits, and social programs afforded their military counterparts. EEOICPA created a new bureaucratic subject: the "atomic weapons employee," whose sacrifice in the service of the nuclear arsenal now officially merited compensation. "Those who fell ill from occupational illnesses," Richardson argued, "were as much casualties of war as those who fell in Europe, in the Pacific, in Korea, in Vietnam or in Desert Storm."[11]

Legislators characterized the law as an effort to restore the status quo ante, repairing both the bodies and the trust that had been broken. By passing EEOCIPA, they promised not only financial restitution and medical care but a restoration of "wholeness" to the nation's nuclear workforce.[12] "We were acknowledging they had been injured," senator Mark Udall said. "We were acknowledging they were heroes ... [and] announcing that we were going to, as the United States, make them whole" again.[13]

My own introduction to EEOICPA was via motorboat in 2009 while traveling the Hanford Reach of the Columbia River. We were a party of five: two worker advocates who ran a nonprofit organization called Hanford Challenge, a ProPublica reporter writing a story about occupational exposures on site, a Hanford worker named Jay who had been forced to retire as a result of such exposures, and a young graduate student (me).[14]

We launched the boat midmorning, Jay steering us to a steady current at the center of the channel and then silencing the engine so we could drift at its pace. The air was comfortably warm, the river smooth and sky blue, embellished here and there with mirrored clouds. We spotted a lanky coyote trotting behind a faded "WARNING: Hazardous Area, Do Not Enter" sign and watched a pod of white pelicans mingle beside an old wastewater

Hanford's B Reactor, 2017. Photo by author.

pump. We startled a small fawn and its mother who stood frozen in place on the shore, their eyes carefully tracking our movement.

But mostly, we stared at the reactors: nine hulking geometric forms that, together, had fueled nearly half a century of war. Five had been cocooned (reduced to their core structure, reroofed, and welded shut), and three were undergoing additional waste management before they too would be sealed. One of the nine would remain open, however, preserved in its original state as a public monument to U.S. nuclear history.[15]

It had been more than two years since my last visit to the site. In the interim, I had moved to California and started a PhD program in which I proposed a dissertation project about Hanford's cleanup. Now I was in Washington for the summer, volunteering full-time at Hanford Challenge in an effort to reacquaint myself with issues on the ground. The organization provided advocacy and support for workers who raised concerns about occupational and environmental hazards but were being ignored and/or intimidated by management. Some were navigating the complex maze of dose reconstruction and compensation. Others had filed whistleblower claims. All were struggling to reckon with problems that DOE and its contractors had refused to recognize.[16]

I was drawn to Hanford Challenge for its focus on worker health and safety. Before that summer, I had spent little time with the people who actually performed the daily labor of cleanup. I had never considered the embodied realities of tank waste sampling or the sweaty heat of coveralls in the desert sun. I had never imagined a radioactive room from the perspective of the person surveying it, never pondered the subtle anxiety of a broken safety shower or a piece of protective gear that didn't quite fit.

At Hanford Challenge, I had long conversations about these and other mundane politics of exposure. I attended a support group for workers who had inhaled toxic waste vapors and met whistleblowers who had lost both their health and their jobs. I heard stories about resistance and retaliation, about missing dose records, about the risk of speaking up. "You know that old fable about the emperor's new clothes?" one person said by way of example at the organization's small conference table. "He hires these tailors to sew him the finest suit, but because they were con men, they wove the suit out of air and told him the material was so fine he couldn't feel it, but, of course, he was naked. And the emperor is so proud of his new clothes that he decides to parade, you know, before the townspeople. So, he goes out totally naked in his new suit and the townspeople are like, 'oh it's beautiful!' except for one little boy who shouts, 'The emperor has no clothes!'"

"In the fable, everyone laughs, saying, 'finally someone said it,' and the emperor is embarrassed and runs back to the palace and is made a fool." But power is rarely so easily dismantled, he warned. "Of course, in the real story, the emperor turns to one of his guards and says, 'take care of that.' And the little boy is speared in front of the whole crowd and his dead body is tossed into the river and the people continue to applaud." The fable, he explained, was an allegory for Hanford's "chilled work environment." Punishing whistleblowers was "a very effective method of silencing other employees." It made workers "so afraid to raise concerns, that they aren't even willing to say they're afraid."[17]

That summer changed the way I thought about cleanup. I could no longer watch DOE presentations without hearing those stories in my mind, unsettling agency assurances of safety and progress. I remembered the technician who told me about being fired after they reported a broken waste transfer line and the tank farm worker pressured to falsify their timecard so management could claim "no lost days" to injury. I recalled the operator who had been labeled a "crybaby" lacking the right "Hanford attitude" for raising safety issues at the Plutonium Finishing Plant. "I'm surprised one of the guys hasn't killed him by now," one personnel manager said.[18]

But it was more than the site's broken safety culture; it was also the structural erasures integral to remediation itself. It was the blind faith that land use regulations would last for millennia. It was the risk assessments that didn't consider the synergistic effects of radioactive and chemical toxicants.[19] It was Reference Person's normative body. And it was the fact that someone like Jay, who had spent his entire career at Hanford, would have to fight to be recognized at all.

As we drifted past the reactors where he had worked for years, Jay told us about his exposure. It began with a strange smell in the tank farms, followed by a nausea so extreme he ended up in the hospital. His eyes swelled shut. He struggled to breathe. He had trouble organizing his thoughts. And, though he returned to work soon after, his symptoms persisted: he was easily winded, brain foggy, heart racing to the point he worried he would pass out.

Jay was eventually diagnosed with a form of occupational asthma known as RADS (reactive airways dysfunction syndrome) as well as peripheral neuropathy and a degenerative neurological disorder called toxic encephalopathy. "I ended up on permanent disability from the exposure I took," he said. The tank farm contractor didn't have any data about the vapors he had inhaled (it wasn't common practice to collect such information at the time), so "they tried to blame my symptoms on smoke in the air and things like that." Jay understood this lack of data as an intentional strategy to undermine workers' claims. "If they know anything, then they can't play the stupid game," he added grimly. "The stupid game is huge out there."

It took three and a half years for Jay to receive compensation and medical benefits through EEOICPA. "You know, I went out there to do a good job, thinking that if a person gets hurt, they would try to help out," he said, his face turned toward the reactors on our right. "Nope. It's a total fight."

Jay's story was one of many I heard that summer about how exposure and illness came to "count" at Hanford. In the years that followed, I would hear many more. And with time, I would notice a familiar contradiction at the heart of EEOICPA's reparative promise. Just as the BEAR report had explained four decades earlier, safety did not mean total protection, but rather an "acceptable risk" of bodily impact. While the assistance that EEOICPA offered did help workers, it did not unmake these foundational logics. Quite the opposite: it normalized exposure as an unfortunate, yet necessary, part of modern life and work.[20]

RESOLUTION

Feminist scholar Donna Haraway argues that "struggles over what will count as rational accounts of the world are struggles over *how* to see."[21] Rationality is a distinct practice of visualization, she writes: "How to see? Where to see from? What limits to vision? What to see for? Whom to see with? Who gets to have more than one point of view? Who gets blinkered? Who wears blinkers? Who interprets the visual field?"[22] Seeing, by its very nature in other words, is both powerfully situated and partial. It is made and unmade in the asking.

EEOICPA faced several distinct challenges when it came to visualizing the workers it promised to make whole. First, the weapons complex comprised more than one hundred sites as well as a variety of government agencies and independent contractors, many with spotty record-keeping practices. Second, permissible dose had been interpreted and enforced unevenly throughout time and space, making it difficult to concretize its effects. Third, it was impossible to say with certainty whether an individual's illness was the result of their nuclear work. How, then, to compensate for injuries that had long been structurally invisible?

Though Congress's JCAE met to address this very thing in 1959, it found that it could not overcome the issue of indeterminate causation.[23] For, it argued, if radiogenic cancers were indistinguishable from nonradiogenic ones, how would a workers' compensation program even function?

The following year, the JCAE's Special Subcommittee on Radiation held seven days of public hearings entitled, *Radiation Protection Criteria and Standards: Their Basis and Use*. Focused on the structural administration of nuclear safety (rather than on compensation for its effects), the hearings considered how best to regulate radiation protection through the auspices of government. Echoing the BEAR report, witnesses and congressional representatives spoke openly about the political and economic necessity of exposure. Indeed, the official hearing summary made clear that radiation protection was not designed to prevent harm altogether: "None of this information contemplates absolute protection. Radioactive materials will be released into the environment. Occupational exposures will occur regularly. Accidents will occur."[24]

Rather, the purpose of the hearings was to determine how to identify and implement an acceptable level of risk.[25] As the representative from the U.S. Naval Radiological Defense Laboratory testified, "We cannot insist on . . . zero risk in the development of new industry which contributes immensely to man's well-being, wealth, and power over his environment. Some degree of biological effect is associated with any exposure to ionizing radiation, and this effect must be accepted as inevitable. Since we don't know that these effects can be completely recovered from, we have to fall back on an arbitrary decision about how much we will put up with; i.e., what is 'acceptable' or 'permissible'—not a scientific finding, but an administrative decision."[26]

Most speakers agreed that exposure should only be considered acceptable when the benefits outweighed the risks. However, translating this general philosophy into a specific numerical value presented significant challenges. As Hanford scientist Jack Healy put it, "We would like to strike a balance in which the maximum benefits are obtained through use of radiation with the minimum harm. [However], we cannot accurately define the risks nor can we accurately define the benefits to all people. Even if we could define these factors, there would still be controversy as to the proper level at which the balance should be taken. . . . Ultimately, our problem resolves itself into the broad area of overall social risk and progress. . . . How do we strike a proper balance between the interests of the individual and the interests of the Nation?"[27]

So too, implementing nuclear safety standards was further complicated by the material realities of radiation itself. First, the human body could not sense ionizing radiation, and its effects often took decades to become apparent. Second, guidelines for radiation protection permitted a different level of dose for each part of the body, allowing more exposure to the skin, hands, forearms, feet, shins, and calves than to the lenses of the eyes and the critical organs. Third, there were multiple types of ionizing radiation associated with nuclear production, each with a different capacity for penetrating the body and therefore different requirements for monitoring and protection.[28] Fourth, each radionuclide had its own unique biological signature. Some, like strontium-90, tended to deposit in bone, while others preferred thyroid tissue or muscle. In addition, radionuclides varied in the amount of time they remained active in the body, and they could behave differently when combined with the chemicals that were also associated with nuclear processing. Finally, calculating exposure required close attention to time—recording how long each part of the anatomy was exposed to each type of ionizing radiation and then adjusting for relative risk accordingly. The result of all of this, Hanford's chief radiation health scientist Herbert Parker testified, was that "man gets so subdivided between time, space, radiation types, and radionuclides that the basic integrating sense of standards is lost."[29]

Thus, though the hearings were convened to formalize administrative standards for radiation protection, much of the committee's time was spent addressing the impossibilities of such an endeavor. As the official summary stated, "Testimony presented at the hearings indicated that the major difficulty in translating [dose] recommendations into legal status apparently lies in the fact that such use of the recommendations is not compatible with the philosophy and rationale of the recommendations themselves."[30] Not only was it difficult to accurately calculate dose, but such calculations would fail to incorporate the necessary political, economic, and/or technical context of each exposure event. In other words, such an explicit standard would be unable to account for the fact that "acceptable exposure" was a flexible and, more importantly, a *social* determination informed by the entwined imperatives of war and industry.

Years later, the federal government adopted the principle that radiogenic exposures be as low as reasonably achievable (ALARA), "taking into account the state of technology, and the economics of improvements in relation to benefits to public health and safety, and other societal and socioeconomic considerations, and in relation to the utilization of atomic energy in the public interest."[31] ALARA's objective was to expand the margin of safety for radiological workers by encouraging nuclear facilities to lower their occupational dose rates below the maximum limits allowed by law. As such, it was designed to give workers additional protection in radioactive space: "an enhanced safety factor for what are already considered to be safe annual doses for radiation workers."[32]

In theory, ALARA provided a procedural framework for addressing the uncertainties of radiation protection. It recognized that exposure produced a spectrum of hazard and acknowledged the embedded cost-benefit calculus of nuclear safety. Perhaps most importantly, it positioned federal dose limits as the uppermost boundaries of harm and then rendered everything below that safer by comparison. As such, it produced some exposures as more protective and thus more reasonable than others.

So too, ALARA framed dose as a rational choice: the result of thoughtful, daily decisions by workers and managers.[33] As Caroline Schieber and Christian Thezee of the Centre d'étude sur l'Evaluation de la Protection dans le domaine Nucléaire (roughly, Center for the Study of Nuclear Protection) put it, successful ALARA implementation means engendering a "state of mind" among the workforce and creating "a shared radiological protection culture" with a "common language and system of values" about exposure.[34] Only then, they argue, "will it be possible to achieve a socially acceptable compromise between the various risks and the resources and means allocated for their management."[35]

Critically, though ALARA requires that workers obey procedural and managerial direction, it also makes personal judgment a natural basis for radiation protection. DOE's *Radiological Worker Training* handbook, for example, describes the decision to engage in nuclear labor as integral to exposure's rationalization. "Acceptance of a risk is a personal matter [that] requires a good deal of informed judgment," it explains. "The individual radiological worker is ultimately responsible for maintaining his/her dose ALARA."[36]

This logic is especially striking in personal injury cases, where the law requires a concrete definition of "unnecessary" or "excessive" exposure in order to determine culpability. Though a regulatory principle, ALARA is not considered a legal "standard of care" in toxic tort litigation.[37] As lawyers Donald Jose and David Wiedis write in *Radiological Safety Officer Magazine*,

If ALARA were the standard of care, every exposure, no matter how small, could potentially make the [nuclear] licensee liable for negligence since every exposure could be analyzed with the benefit of hindsight and in most instances, it would be 'possible' to have reduced the exposure. This would undermine the very stability that the regulations were designed to provide because licensees would be held liable for allowing a dose that regulations specifically labeled as permissible.[38]

While injured workers can claim that doses above federal maximums have caused excessive harm, ALARA's risk calculations do not hold the same legal weight. Instead, by showing up to work, employees are expected to have knowingly accepted the terms of permissible dose. "All exposures are in a sense unnecessary to [the worker] because he could simply elect to not be a nuclear worker," Jose and Wiedis argue. "Once he elects to be a nuclear worker, he consents to receive an exposure within the federal numerical limits, so that exposure cannot then be called unnecessary. It is a necessary part of the job he has chosen to pursue."[39]

Thus, ALARA formalizes a flexible notion of protection while avoiding its legal liabilities. It does so in the language of reason—creating spreadsheets, compiling data, and making workers "ultimately responsible" for maintaining a level of risk that they personally consider acceptable. Indeed, it frames labor itself as an implicit form of consent, presenting "exposure within federal numerical limits" as "a necessary part of the job."

LIVING IN DOSE

Echoing Cold War civil defense narratives of decades past, DOE's *Radiological Worker Training* handbook uses probabilistic comparisons to instruct workers in risk interpretation. "The risk of cancer induction from radiation exposure can be put into perspective," it explains. For example, "the current rate of cancer death among Americans is 20%. Taken from a personal perspective, each of us has about 20 chances in 100 of dying of cancer. A radiological worker who receives 25,000 mrem over a working life increases his/her chance of dying of cancer by 1%, or has about 21 chances in 100 of dying of cancer. A 25,000 mrem dose is a fairly large dose over the course of a working lifetime. The average annual dose to

DOE workers is less than 100 mrem, which leads to a working lifetime dose (40 years assumed) of no more than 4,000 mrem."[40] (Here, DOE is assuming that ALARA practices will maintain occupational exposures below federal limits. It should be noted, however, that permissible dose for DOE workers is actually 5,000 mrem per year, meaning that a worker could receive a lifetime dose of 200,000 mrem and still be considered reasonably exposed.)

The handbook also includes a table designed to help workers "put potential risks into perspective when compared to other occupations and daily activities."[41] Juxtaposing a range of industries and personal practices, it outlines relative decreases in life expectancy from things like smoking cigarettes, working in agriculture, and being struck by lightning. At a reduced life expectancy of just fifty-one days, for example, it positions occupational exposure as safer than spending time at home (where, it estimates, accidents decrease the average lifespan by seventy-four days).

"The estimated risk associated with occupational radiation dose is much less than some risks widely accepted by society," the handbook continues. "The risk of work in a radiation environment is considered within the normal occupational risk tolerance by national and international scientific groups. However, the acceptance of risk is an individual matter and is best made with accurate information. . . . It is hoped that understanding radiation risk and risk in general will help you to develop an informed and healthy respect for radiation, and that your understanding will eliminate excessive fear."[42]

These training materials not only normalize exposure, they also ignore occupational power dynamics that make it difficult to simply accept or decline risks at will. So too, they outline a set of expectations for rational decision-making: choices that do not reflect these terms may be considered suspect, the result of "excessive fear" and/or misinformation.[43]

Ultimately, reasoning exposure means accepting that safety can coexist with risk. As a longtime Hanford worker named Stephanie told me, "We know how to protect ourselves, how to be safe. Now, we still do things, you have to take risks. . . . We're out there monitoring our exposure, monitoring our dose, but to do this you have to have a certain buy in of understanding. I do work in the nuclear industry and there are some times that we're going to have proceed forth with a certain amount of risk. And that's just life. That's just how it goes. We do that every single day anyway. And I realize that the catastrophes can be enormous, I fully understand that, but we have a pretty good handle on it."[44]

Being a nuclear worker also means developing an embodied knowledge of the invisible, calibrating physical movements according to radiation sensors and procedural directives in order to avoid unnecessary dose. "It's not just cutting up cheese and turkey, for crying out loud!" Stephanie continued. "If I'm going into some nasty crap, I gotta have a decent work plan, we gotta know what we're doing. If not, your risk is going to shoot up astronomically and you're going to have a lot more uh-ohs out there."[45] Or, as another longtime worker named Jenny put it, "We have to be in constant control."[46]

In her study of radiation protection at Canadian nuclear power plants, historian Joy Parr calls such practices "embodying the insensible."[47] By perceiving exposure through technological proxy, she writes, nuclear workers learn to experience instrumental readings "as sensation, as a form of tacit knowledge with the same credibility as touch, taste, or smell."[48] As Hanford operations supervisor Wakefield "Wakie" Wright described it, "The radiation danger was always with us. We were taught to obey our instrumentation, like flying an airplane. If your instrument says you are flying straight and you think you are flying upside down, you better think you are flying straight."[49]

Measuring protection through bodily quotas and accumulated dose also means fine-tuning oneself to the statistical reasoning of acceptable risk. "You're regulated by the procedure," Arthur, a retired Hanford engineer, told me. "It just gets into your head: how you do your work and how you live your life. It has to get into your head."[50]

This embodied practice requires years of experience working with nuclear materials. "You have to do the work to really understand what it's like," Stephanie said. "New people don't understand how the systems interact, don't understand what specific things mean, and don't understand the interactions." Jenny agreed: "This is stuff you learn over time, over years of knowing history and knowing what you're dealing with. These new kids think they know everything, and it's like, yeah, you're twenty-two years old, you're not really trained. You just spent six months in a classroom doing calculations. Well, this is a high-rad area, stuff you don't have any clue about."[51]

For Jenny, two decades of on-the-job experience had afforded her an increased measure of safety. "We have a certain way that we work around here and we do it for a reason: to keep people safe," she told me. "I want to go home safe. I want to go home to my kids at the end of the day." Still, she recognized that her tenure had also increased the likelihood that she would ultimately develop an occupational illness. "Well, you know, with radioactivity, you get a chronic exposure to that eventually," she continued. "I mean, it's a cancer-causing agent. The scary part is I've been [here] a long time. And you know, I'm at that point, that latency period, you know. About 20 years. You know, where stuff starts happening or people start getting sick or things just aren't right. And so that worries me, it's like, gosh. But you know, the damage has already been done. I mean, I suppose I could quit now and there would be no more damage for me from this day forward, but . . ."; her voice trailed off, leaving the sentence unfinished.

AT LEAST LIKELY AS NOT

EEOICPA's compensation process simultaneously seeks to redress and reproduce the structural logics of reasonable exposure. In particular, it presents quantitative dose reconstruction and probability of causation (PC) as evidentiary frames, relying on statistical modeling to manage systemic uncertainties. Workers who can prove that their illnesses are "at least likely as not" (50% or greater PC) related to their employment are eligible for financial and medical benefits.[52]

While EEOICPA is straightforward in theory, its primary architect David Michaels admitted that the program is anything but in practice. "Historians who look back on this legislation will shake their heads in wonder at the strange beast with all the weird appendages."[53] There are separate procedures for radiogenic versus chemical exposures and convoluted work-arounds for missing dose records. There are opaque computer models and controversial assumptions about disease rates and exposure pathways.[54]

Though the process is intended to "yield consistent, scientifically-informed causation determinations," workers and their advocates have reported that it feels more like an adversarial black box.[55] "They have this 'magical calculation' [for dose reconstruction] . . . which we are not allowed to look at, nor are we allowed to challenge," an advocate and former worker named Viv told me.[56] "We can only challenge what goes into the calculation, like missing data or erroneously applied data. We're never told how the calculation works specifically and, under the law, we are not allowed to challenge it. Is that idiotic? Yes. Do we think it's fair? No. Can we do anything about it? No."[57]

"It's a mystery," an advocate named Maggie agreed.[58] "I went to a workshop they sponsored about it [dose reconstruction], but honestly, I can't even tell you what I learned." At one point, workshop organizers "gave us five dice and six nickels to do a statistical thing," but it was all very abstract. The process was, as one union representative put it, "a Rube Goldberg device of the finest complexity."[59]

So too, EEOICPA places the burden of proof on workers to establish probability of causation. This means that claimants have to negotiate not only their own illnesses but also the program's arcane bureaucracy. "They're absolutely exhausted," an advocate named Clara told me, describing both the workers and the family members who care for them. When someone is sick or dying, "you're lucky if you can just take a shower or get a night's sleep . . . and you're talking about an elderly community, an elderly ill community who, unfortunately, if they don't have a good advocate under this program, the chances of them expiring prior to ever getting compensated are great."[60] In addition, Viv pointed out, "you are mostly dealing with men over 60 years of age and up, and when they become incapacitated, their wife becomes their care giver." She worried that these gender dynamics could influence requests for in-home assistance, allowing officials to deny claims by saying, "Well, they have a wife."

Many of the advocates I spoke with, most of whom worked as independent consultants, had learned to navigate EEOICPA as claimants themselves or as members of a claimant's family. "I never intended to do this as a job," Viv told me. "People found out I was working on my own EEOICPA claim and it's all been word of mouth." Recounting her husband's experience seeking compensation, Maggie explained, "The reason the advocates' fees are so low is because we aren't ambulance chasers. . . . The process is so complex especially if you are sick and you can't understand what they are saying because you are in pain. You need someone to help you."

Dose reconstruction, for example, requires documentation about one's personal exposure history, but such records are often nonexistent or lacking crucial details. Sometimes monitoring data is lost in the transition between site contractors or misreported in the name of national security.[61] Other times, it isn't recorded at all. Jenny told me about the day she received six hundred millirem while pregnant—one hundred millirem more than the permissible dose for her entire pregnancy. "They never assigned me the dose because then I would have been overexposed for a pregnant worker," she said in frustration. "They said that there was probably just something wrong with my dosimeter, that there was a pinhole—a microscopic pinhole in my dosimeter that nobody could see—that would cause it to malfunction. I said, well that's crap. I'm not buying that. And I've gone back and said, hey, you know, I'm curious about this. You know, I was pregnant! And they said, no, they have no record of it. None whatsoever."

In addition, dosimetry practices on the job have the potential to undermine compensation claims. When it comes to dose reconstruction, advocate and former worker Dan explained that it matters where dosimeters are placed on the body.[62] He told me about a claimant seeking compensation for a debilitating thyroid condition who had been denied because he could not prove that his thyroid had been overexposed. "Your dosimeter, you were told, had to be in the center of your chest, and that's well understood if you are protecting the heart and lung," he said. "But in this person's situation, the [thyroid] dose was never recorded because there wasn't any dosimetry in the center of his neck. So, then you submit the paperwork and the people [assessing the claim] say, 'well, the reading was zero' or 'we don't have any record of this.' Because there wasn't a dosimeter at that place, the dose was unmeasured. But, no reading does not equal zero exposure."

Viv also described how the logic of safety as administered through dose can be undermined in the recording process. "When I would dress out, I was told that my [dosimetry] badge went on the inside of my coveralls," she told me.[63] My pencil dosimeter, if it was issued, was in my pocket on the outside. But my badge that had the secondary dose on it, was *inside* my coveralls! . . . That was the way you dressed out because they didn't want you crapping up your badge.[64] 'Don't crap up your badge,' they said.

What's more important here, a two-dollar badge or your life? The badge!" She laughed sarcastically.

These erasures matter in dose reconstruction because, as Maggie put it, "How do you submit as evidence something that wasn't recorded?" However, missing data is but one of many uncertainties built into the compensation process. EEOICPA derives its risk metrics from Hiroshima and Nagasaki survivors, for example—extrapolating from the single, high-level dose of a bomb blast to assess the chronic, daily effects of nuclear work. This is controversial not only because they are unique exposure contexts, but because the atomic bomb survivor studies are themselves a source of uncertainty.[65]

Furthermore, dose reconstruction refracts bodies and contaminants into normative input parameters. It reduces workers to risk tables and itemized lists of organs, imagining radionuclides as finite actors with linear exposure pathways. As such, geographer Shiloh Krupar writes, EEOICPA fixes "processes and ecologies of the work environment in place and time," eliding the bomb's "materially linked but differentially inhabited risks."[66]

Krupar argues that uncertainty is not only a structural reality of the compensation process but a means for denying financial and medical benefits as well. "EEOICPA's methods fold that which is ontologically indeterminate *with* historically produced voids of information *into supposedly knowable, calculable, epistemic uncertainties*—into 'certain uncertainties,'" she writes (italics in the original).[67] This lack of evidence often functions as evidence in and of itself where "the absence of information is produced as all there is to know."[68]

Because they cannot alter EEOICPA's statistical models, workers and advocates do their best to plug its epistemic holes. They spend hours poring over medical records and academic journal articles, searching for anything that might get them above the 50 percent PC threshold. If a primary cancer isn't covered, Clara told me, she searches the claimant's records for related conditions that might be eligible instead. "I'll ask, did that cancer go to the bone, kidney, or lung? I'll dig through thousands of pages of medical looking for any sort of pathology. I'll read the EKGs, I'll read the CTs—anything I've got—because that could be the difference between a compensable and a non-compensable case."

Perhaps the simplest route through the compensation process is via EEOICPA's Special Exposure Cohort (SEC) status, which recognizes particular buildings, sites, and time periods when and where records are so poor that injured workers have "presumptive eligibility."[69] SEC status was designed for contexts in which dose reconstruction is "scientifically infeasible or morally unnecessary" and allows claimants who have worked for at least 250 days during an SEC time period (and have one of twenty-two covered cancers) to automatically receive compensation.[70] If no SEC exists for their particular occupational history, workers can petition to establish one. However, as Viv noted, the process often takes years and "people are sick ... most of them don't have the wherewithal to do it."

Clara filed the first SEC petition through EEOICPA after losing her parents to cancer (both of whom had worked on the Manhattan Project). "The reason I filed was because in my research for my father's claim ... I uncovered documents about record manipulation. They [managers at the facility where her parents worked] were altering records due to liability concerns. One of the documents actually stated, 'this poses a unique opportunity to conduct a clinical experiment on such a widely exposed population.' ... So, when I filed a petition for those workers, I had that proof of record destruction, record manipulation, and lack of monitoring. I attached all of that to my petition, and it was approved."

In the years since then, claimants and advocates have filed hundreds more petitions across the nuclear complex. As of 2021, more than two decades after EEOICPA's passage, SEC classes were available at seventy-seven facilities, and about 70 percent of compensation claims for radiogenic cancer fell within an SEC.[71] Each petition sought to expand the terms of visibility, making it easier to (officially) *see* the bomb's embodied effects. Collectively, they also unsettled the notion of "special exposure" by identifying the ways that missing data and unrecorded doses were "not the exception, but the rule within the history of the U.S. nuclear complex."[72]

Still, advocates argue that SECs produce their own erasures as well. The 250-day requirement, for example, fails to account for temporary or contract workers who only spend a few months on the job. Coverage is also primarily limited to a particular list of cancers, ignoring other consequential conditions. In addition, there are no legislative remedies for family members exposed to contaminants that workers may have brought home on their clothes, nor does EEOICPA recognize possible mutagenic effects in workers' children. Further, the act only offers compensation for site employees, excluding downwind communities whose exposures are not occupational in nature.[73]

EEOICPA's makers described it as a rational mechanism for remedying the unequal social relations of nuclear production. They imagined that workers would submit their wounded bodies as evidence that they had not received adequate protection, and that compensation would then be distributed (or not) according to the logic of the marketplace that gave their labor value.[74] Injury, the theory went, would be "'undone' through the monetary award that [would] in a rough sense 'buy back' what it [had] taken."[75]

However, as Elaine Scarry has argued, "compensation is only a mimetic rather than an actual undoing."[76] While the financial and medical assistance that EEOICPA offers may prolong and improve workers' lives, it does little to address the fact that the nuclear state continues to produce the injuries it adjudicates.[77] Instead, compensation rearticulates the broader social logic of permissible dose that makes "safe" synonymous with "safe enough."

When hundreds of Hanford workers and their families gave testimony in the months before EEOICPA's passage, their stories revealed more than a shared history of illness and injury. At stake was the meaning and value of their sacrifice—whether their wounds carried social, political, and economic weight.[78] As one worker put it, "There's a lot of us who worked hard for a lot of years. We're not looking for a free lunch. We're not interested in a big money claim. But you don't want to be left with nothing, with no health to go get another job. We just want enough to get by and to live a dignified life for what's left of it."[78] Another made the comparison more direct, pointing to the ongoing financial investment that the federal government was making in Hanford's cleanup. "The Department of Energy is spending over six billion dollars to clean up [its] contaminated waste. A billion of that is coming to the Hanford Site.[80] We're only asking that the Hanford workers get treated just half as well as the dirt."[81]

5 Trespassing

We entered the Columbia on a warm, windless morning in early October 2017, searching for radioactive escape. The Hanford Reach was calm and cerulean, softly rippled like old glass, its placid surface masking a powerful current below. Though only knee-deep along the shore, the water was surprisingly cold through my waders and, for a moment, I wondered if they had sprung a leak. However, as I scanned the material for holes, I realized that the river hadn't reached my skin; it had merely applied a distinct, swirling pressure. What I felt was its movement, the frigid curl of eddies around my legs, memories from a childhood spent in water just like this.

It was strange to be standing on the Hanford side of the river. I was there with permission—tagging along with Washington Department of Health (DOH) staff as they surveyed the site's fluvial boundary—but stepping onto dry ground still felt illicit somehow. It had all been so easy. There was no fence. No one there to stop us. The act's very simplicity made it that much more surreal.

Our plan for the day was straightforward: we would travel the length of the Reach by boat, stopping every few miles to replace the dosimeters DOH had installed in the spring and collect the used ones for laboratory analysis. The devices were small, about the size of a silver dollar, zip-tied to old posts and tree branches, where they had spent six months absorbing gamma radiation.[1] Retrieving them would require bushwhacking through willows and dogbane, wild tarragon and cheatgrass, thick bunches of rabbitbrush with its fall yellows and gray-greens. Some would be easy to find while others would take concentrated effort, parsing the two seasons of vegetation that had swallowed them whole.

It turned out to be pleasant work. I enjoyed the scavenger hunt-esque challenge of our search and the lazy stretches of river travel in between. Our crew leader's wife had packed everyone sack lunches, and we ate them contentedly on the boat, chatting about reactors and fishing and birds. Then we headed shoreward again, hopping into the river and wading through the brush until someone shouted a triumphant, "Got it!"

Still, I couldn't ignore the way my stomach clenched each time we crossed Hanford's invisible boundary. I felt that breach as a physical sensation: a prickling anxiety at how easy it was to exceed the permissible. On the few occasions we encountered sun-bleached signs warning us to keep out, I laughed uncomfortably at their impotence. *This* was securing the site's border? Its message was nearly indecipherable.

As the day continued, however, I became less interested in Hanford's poorly marked edges than in my own embodied reaction to them. Though my research unsettled the categories that defined this land, I found myself reproducing them each time I stepped onto the site. When my heart quickened with the understanding that I had entered restricted territory, I was responding to the idea that Hanford *could* be bounded. After all, in order to knowingly trespass in a particular space, one first has to recognize its container.

Such recognition is essential to cleanup's very logic. According to federal law, remediation must prevent unreasonable exposure for thousands of years, managing not just Hanford's waste but the future humans who may encounter it. Because most of that waste will remain on-site, this involves a combination of physical remedies (like capping, entombment, and moving contaminated material from one location to another) and institutional controls (like land use restrictions designed to "protect the integrity" of said remedies).[2] It also requires that DOE articulate a vision for human activity after remediation is complete in order to demonstrate that requisite safety standards will be met.

No Trespassing sign along the Hanford Reach, 2017. Photo by author.

As the agency describes it, current waste management activities should "reflect the planned future use of the property" and be "appropriately protective of human health and the environment consistent with that use."³ Hanford's "end state," DOE explains, provides "the basis for exposure scenarios developed in baseline risk assessments that help establish acceptable exposure levels for use in developing remedial alternatives."⁴ Cleanup, in other words, exists in tautological relation with its imagined future and the statistical people who inhabit it. These probabilistic figures are at once the ends and means of risk-based remediation, making cleanup possible by performing its conditions of possibility.⁵

In 1999 DOE issued a comprehensive land use plan (CLUP) for the Hanford site detailing its expectations for post-cleanup life. Developed in conversation with tribal governments and a collection of federal, state, and local agencies, the CLUP describes nine possible land-use categories— industrial (exclusive), industrial, agricultural, research and development, high-intensity recreation, low-intensity recreation, conservation (mining and grazing), conservation (mining), and preservation—as well as six alternative visions for the future, each representing a different stakeholder perspective.

DOE's "preferred alternative" designates most of the 220 square miles along the Columbia River for preservation and conservation (mining), with several smaller spaces for high- and low-intensity recreation. It categorizes the Central Plateau—75 square miles containing some of Hanford's most hazardous material—as industrial (exclusive) and industrial to allow for ongoing waste management activities. Finally, it classifies the southwest portion of the site, home to some of the area's laboratory facilities, as a mix of industrial and research and development.[6]

Though the CLUP boasts "cooperation" and "consultation" with stakeholders, it does not offer a shared vision for Hanford's future. Most notably, it describes significant disagreement between DOE and tribal governments about how to interpret treaty-protected activities on Hanford land. These interpretive differences were so great during the design process, in fact, that the parties "agreed to disagree" with the understanding that the issue would be revisited in future conversations.[7] As DOE explained at the time, neither the existence of the CLUP "nor any portion of its contents is intended to have any influence over the resolution of the treaty rights dispute."[8]

Yet because the CLUP's function was to inform future land use, it was unclear how agreeing to disagree would actually work. If DOE did not recognize tribal members' rights to gather traditional foods and medicines, for example, then it was unlikely that cleanup would make such activities possible. Revisiting treaty interpretation after the CLUP had been issued seemed to ignore the procedural logic of remediation itself.

Indeed, when DOE issued supplemental analyses (SAs) of the CLUP in 2008 and 2015, it no longer framed treaty rights as an open question. Instead, it presented its preferred alternative as *the* plan, unmarred by previous interpretive disputes. Although in 2015 the SA noted that "the tribes have expressed interest in consulting with DOE about the potential of returning to some of their traditional practices and lifeways," it described Hanford's future as a fait accompli predetermined by the CLUP. "Any proposed increase in public and tribal access to Columbia River shoreline for recreational activities would be evaluated for consistency with [the]

CLUP land use map," it explained, and "limited public and tribal access is consistent with resource conservation, recreation, and preservation land-use designations."[9] Not only did the SA inappropriately equate public and tribal activities (calling the tribes' use "recreational"), it also bound future populations to DOE's vision for the site.[10] To allow tribes unrestricted access would risk exceeding "acceptable exposure levels" and thus undermine the agency's definition of clean.

Still, even as it claimed categorial fixity in limiting these activities, DOE also described the CLUP as a "living document ... flexible enough to accommodate a wide spectrum of both anticipated and future mission conditions."[11] It argued that a "fundamentally good plan" should be iterative, informed by ongoing "monitoring, data gathering, and analysis for the purposes of 'fine tuning' and improving the plan by Amendment."[12] In that spirit, SAs were framed as opportunities for bureaucratic reflection—a process by which DOE could consider "significant new information or changed circumstances" that might warrant revision.[13]

Despite repeated requests from state agencies and tribal governments, however, DOE has yet to amend the CLUP's land-use designations.[14] Instead, it uses the document's very existence to make the case against amendment, presenting it as evidence that changes remain unnecessary. "DOE has routinely cited the CLUP as definitive guidance for long-term land-use decisions related to cleanup, to the virtual exclusion of other factors," Ken Niles of Oregon's Nuclear Safety Division wrote in 2008. "This has been particularly true for decisions that limit cleanup to something less than an unrestricted use standard. Because the range of options under CERCLA decisions has been constrained by the CLUP, it should not come as a surprise to DOE or readers of the SA that decisions made through the CERCLA process are consistent with the CLUP."[15] In other words, the evaluative context is inherently circular: the CLUP's vision makes the CLUP's vision possible.[16]

In the process, it also reflects a broader power dynamic integral to remediation's regulatory structure. According to federal law under the Comprehensive Environmental Response, Compensation, and Liability Act (CERCLA, also known as the Superfund Act), DOE is responsible for determining the "reasonably anticipated future land use" that will guide risk assessment and cleanup decisions at Hanford. As EPA describes in its

1995 "Land Use Memorandum" (which provided regulatory guidance for the CLUP), such planning assumptions should be informed by conversations with local stakeholders in order to "increase confidence [that] expectations about future land use are, in fact, reasonable."[17] However, EPA also acknowledges that such expectations may not prove "cost effective" or "practicable" upon further investigation. "If this is the case," the memorandum continues, "the remedial action objective may be revised which may result in different, more reasonable land use(s)."[18]

Reasonableness, therefore, is a powerfully ambiguous term—at once abstract and concrete, flexible and determinative. It frames stakeholder input as necessary to understanding future human activity and then rejects that information when it trespasses the limits of budget and practicability (another powerfully ambiguous term). As such, it provides a language and logic for cleanup's administrative contradictions, making and unmaking the meaning of protection.

DOE, for example, employs the National Environmental Policy Act (NEPA) "rule of reason" in addressing the CLUP's uncertainties. This rule states that federal environmental analyses need only describe "reasonably foreseeable" effects of cleanup actions that are "supported by credible scientific evidence" and "not based on pure conjecture."[19] However, because modeling impact in deep time is an inherently speculative exercise, the line between credible and conjectural is often a matter of debate. What makes one future more reasonable or foreseeable than another?

"If there is total uncertainty about the identity of future land owners or the nature of future land uses," the CLUP explains, "then of course, the agency is not required to engage in speculation or contemplation of their future plans."[20] At the same time, it also maintains that *some* speculation is essential to responsible land-use planning. "In the ordinary course of business," it continues, "people do make judgments based on reasonably foreseeable occurrences. It will often be possible to consider the likely purchasers and the development trends in that area or similar areas in recent years; or the likelihood that the land will be used for an energy project, shopping center, subdivision, farm, or factory. The agency has the responsibility to make an informed judgment, and to estimate future impacts on that basis, especially if trends are ascertainable or potential purchasers have made themselves known. The agency cannot ignore these uncertain but probable effects of its decisions."[21]

DOE makes clear, therefore, that land use planning should reflect contemporary development trends. It defines "informed judgment" by near-term industry and real estate markets and "estimate[s] future impacts on that basis." Thus, it excludes futures that do not fit within this frame, positioning them as improbable. As Marlene George, project manager for the Yakama Nation's Environmental Restoration and Waste Management office, put it, "Tribal exposure scenarios are treated as uncertainties rather than being included among the current and future reasonable land use scenarios."[22]

DOE's "Hanford Site Cleanup Completion Framework" reiterates this rationale. Created to "enhance dialogue" between agencies and stakeholders about remediation strategy, it describes both short- and long-term projects as well as "what activities are possible for the site once cleanup is complete."[23] It emphasizes that remediation efforts "will support anticipated future land uses" that have been "bounded by" the CLUP.[24]

The "Completion Framework" also clarifies that, as part of the remedy selection process under CERCLA, DOE often *considers* land uses that have not been sanctioned by the CLUP. For example, it may use a resident farmer or resident tribal member scenario to estimate "potential impacts to specific populations from unexpected exposures."[25] However, it explains that the agency must ultimately select remedial alternatives that will "lead to site activities which are consistent with the reasonable anticipated land use."[26] Such remedies, it continues, are based on "realistic and defensible human health exposure scenarios" informed by the CLUP.[27]

Here again, DOE uses the CERCLA process to make a circular argument about Hanford's future. It employs farming and tribal scenarios to assess potential impacts from "unexpected exposures," but such exposures only qualify as unexpected because they are not part of DOE's preferred alternative for the site. Indeed, the very requirement that remedy selection must be "consistent with reasonable anticipated land use" relies on the same tautology. DOE foresees the future it deems reasonably foreseeable.[28]

Hanford's Long-Term Stewardship (LTS) program has also been tasked with preserving these boundaries after cleanup is complete. Initiated in 2010 under the tagline "Bridge to the Future," LTS is responsible for "ensuring the protectiveness" of "all engineered and institutional controls designed to contain or to prevent exposures to residual contamination and waste."[29] Such activities include surveillance and record keeping, monitoring groundwater plumes and repairing caps, and maintaining containment structures and fence lines in accordance with the site's land-use designations.

As part of its mandate, LTS conducts CERCLA 5-Year Reviews of completed projects to determine whether they remain protective. In particular, it asks:

1. Is the remedy functioning as intended by the decision documents?
2. Are exposure assumptions, toxicity data, cleanup levels, and remedial action objectives used at the time of remedy selection still valid?
3. Has any other information come to light that could call into question the protectiveness of the remedy?[30]

Ostensibly, the review process is an opportunity to revisit cleanup's logic. Its framing questions recognize the contextual nature of remediation, allowing that protectiveness depends on historically situated objectives and assumptions. And in assessing whether a particular remedy is still functioning as intended, LTS may find that it requires a new approach.

Each time the 5-Year Review comes around, it feels like an opening to me. The HAB writes policy advice; the agencies hold public meetings; and stakeholders provide formal comments, analyses, and perspectives from across the region. Thus far, however, these reviews have yet to meaningfully reimagine Hanford's reasonably foreseeable future. Instead, they tend to reiterate the powerful relations at the heart of cleanup, reproducing the statistical lives it was already designed to protect.[31]

Perhaps I should not be surprised by this. Despite its emphasis on self-reflection and revision, LTS also depends on administrative fixity—promising that governmental institutions will endure and have the wherewithal to monitor remedies long into the future. In fact, one of the program's biggest challenges is simply continuing to exist for the lifespan of Hanford's waste.[32] "Here, what we're doing is babysitting trash," a worker once told me. "And when you babysit trash, you're always on the chopping block for funding."[33]

In addition to assessment, then, the 5-Year Review is also invested in demonstrating its own durability. As one remediation specialist put it, "I believe the purpose of the 5-Year Review is affirmation. It's EPA saying 'we're still here, our ICs [institutional controls] are still active, and we think we can make it another five years until the next one.'" The review process is not only about protecting remedial integrity, therefore, but also about resisting the fact that time will inevitably unravel the structures that administer it. "It's 5-Year increments into forever," he said, "to stave off the thought of institutional death."[34]

FRUTONIUM

Managing cleanup's boundaries in deep time is a more-than-human challenge as well. "I mean, think about it," an EPA staff member told me years ago, "animals can't read signs."[35] Thus, LTS is also tasked with regulating what it calls "biotic vectors": the flora and fauna that trespass in and beyond "radiologically controlled" space.[36] As DOE explains in its 2020 annual environmental report, "Each year on the Hanford Site, legacy contamination is spread from environmental conditions. Rodents eat vegetation located in contaminated areas and deposit contaminated feces outside of the contaminated area. Birds build nests and occasionally use materials from contaminated areas, resulting in contamination transfer to uncontaminated areas."[37] Tumbleweeds, with their deep tap roots and propensity for long-distance travel, are particularly adept at redistributing radioactive materials. Hanford workers used to mark the contaminated plants with paint in order to track their movement. "We'd spray them pink to watch and record how far these things went," one former worker recalled. However, when they began turning up in Richland, "People started freaking out, so we quit painting them. But that doesn't mean the radioactivity stopped."[38]

Stories like these are a familiar part of the cleanup landscape. Some circulate as jokes and snarky asides, oblique references to Hanford's radioactive excess. Once, for example, after arriving late at a HAB meeting, I explained that I'd had to pull over on the side of the road and disentangle tumbleweed from the grill of my car. When I sat down at the table, a fellow board member nudged me with his elbow and whispered, "Was it pink?" We both laughed sardonically.

Other tales have become legend with decades of retelling, their details softening with time. There is Norm Buske's jam, made with mulberries he

A tumbleweed waits at one of Hanford's entrance gates, 2017. Photo by author.

picked near the N reactor and sent to Washington's governor as evidence of site hazards.[39] There is the worker who set off radiation alarms after eating oysters from Willapa Bay (more than one hundred miles downstream of Hanford).[40] There are the irradiated alligators that escaped from the site's Experimental Animal Farm in the 1960s, slipping into the Columbia River unseen. Project scientists recaptured two of them and a local fisherman caught the third, displaying it in a nearby sporting goods shop until Hanford officials confiscated it. "The last two were never found," a retired worker warned a group of us on our way to the B reactor in 2017. "Keep a look out," he winked.[41]

The story I return to most often, however, is the "fruit fly incident" of 1998.[42] It began as a mystery, a health physics technician (HPT) named Tracy told me at her kitchen table as we sipped hot mugs of cherry tea.[43] In September of that year, her fellow HPTs had identified a strange pattern of contamination in a mobile office trailer that workers were using as a lunch room. First, they noticed that the trailer's light switch was radioactive, followed by a small kitchen knife and cutting board, leading them to believe that the source was someone's contaminated hand. When they found radioactive chewing tobacco in the trailer's garbage bin, they worried that the individual may have been internally contaminated as well. Later that night, when a worker reported a pair of contaminated socks in their home, the question appeared to be answered.

However, bioassays of the worker's urine did not show elevated levels of radioactivity.[44] Instead, an HPT identified the source the following day when she noticed a speck of contamination in the same trailer "fly away."[45] Perplexed, she called her partner and together they repeated the exercise, realizing that fruit flies—a previously unidentified vector at Hanford— were to blame.

In the weeks that followed, DOE offered multiple hypotheses for the flies' radioactivity before pinpointing a waste diversion pit that had recently been sprayed with a monosaccharide-based fixative. The insects had been drawn to the sugary-sweet substance, laid eggs in the waste, and their progeny had spread contamination across the site.[46] Subsequent surveys detected their tiny footprints in a bathroom, the ironworker shop, the canister storage building, the clothing of at least three workers, several dumpsters, and eventually the city landfill.[47]

Evoking the plot of *Them!*, a 1954 horror film about mutant killer ants, the incident made for easy headlines: "Hanford's Nuke Site Produces 'Hot' Bugs," "Bugs May Be Spreading Radiation at Nuke Plant," "Radiation Bugging Hanford."[48] The local *Tri-City Herald* published its first article after an anonymous tip, and the story spread rapidly across the country.

DOE management leapt into action, struggling to contain both the flies and the meaning of their escape. It established a situation room near downtown Richland and installed fly traps across the site, spraying affected areas with the insecticide Malathion. It created a ten-acre buffer around the trailer and diversion pit; tested urine from more than one hundred workers; and sent HPTs like Tracy to the city dump, where they spent weeks surveying trash (the crew later made T-shirts to mark the event, featuring a manic-looking cartoon fly labeled "Frutonium-98").[49]

In the end, more than two hundred tons of garbage were returned to Hanford and buried in a low-level waste facility on-site. The entire incident cost about $2 million.[50] However, despite its rigorous, expensive, and very public response, DOE maintained that the insects had never presented a serious risk. As Bob Shoup, vice president of environmental health, safety, and quality for site contractor Fluor Daniel Hanford, explained, the highest readings were between 8 and 12 millirads—approximately "equivalent to a dental X-ray."[51] Regardless, "we have zero tolerance for any contamination spreads," he assured the public.[52] "Any contamination outside of a control zone is of concern because it *isn't supposed to happen*. . . . Even though the level of contamination was very small, any contamination outside of controlled radiation areas is unacceptable" (italics added).[53]

Shoup made clear, in other words, that this multi-million-dollar response had more to do with the sanctity of the boundary than the fruit flies themselves. By apprehending every single insect, drawing detailed maps of their movement, and sending workers to pick through Richland's trash, DOE sought to demonstrate that containment was possible. It used the flies as a proxy for Hanford's waste, noting the breach as an aberration that nonetheless reproduced the line nuclear materials were not "supposed to" cross within normal operating procedure.

As such, the incident pointed to a broader politics of trespassing in the atomic age. Through public displays of border crossing and capture, it emphasized tactical divisions between contaminated and uncontaminated space, narrating characters that *could* be contained.[54] However, like fallout, it also described a world already altered by radioactivity, disrupting the normative boundaries of nuclear security.[55]

Soon after, the National Research Council (NRC) issued a report that argued such breaches were not aberrant but integral to long-term waste management. Given the "irreducible uncertainty" of deep time, it recommended that DOE *assume* its infrastructures would fail "and that much of our current knowledge . . . may eventually be proven wrong."[56] In practical terms, this would require adopting "a precautionary approach . . . one that is self-consciously risk averse and therefore takes remedial actions even when harm is not clearly demonstrated" by agency models.[57] It would mean selecting state-of-the-art remedies while acknowledging their limitations and designing to "facilitate possible re-remediation" in the future.[58] Perhaps most importantly, managing waste would mean managing administrative hubris—not only recognizing cleanup's fallibility, but planning for it.[59]

In its emphasis on precaution—preventing harm that exceeds agency imaginations—the NRC's report feels like a balm. It also feels far removed from the current political climate at Hanford, where infrastructural failure remains a somewhat taboo subject.[60] Indeed, many years after its publication, one of the report's authors told me that he continued to find this dynamic frustrating. "Our basic message was you don't know enough. . . . [Y]ou should do this in a staged way and admit what you don't know. But no, Department of Energy is never going to do that."[61]

However, as much as I appreciate its emphasis on humility in the face of uncertainty, I wish the report spoke more explicitly about power. Planning for fallibility should do more than recognize cleanup's "unknown unknowns"; it should also reckon with the social relations of failure and success.[62] It should trace how recognition is unevenly distributed in documents like the CLUP, how some realities become more "knowable" than others. It should name the settler colonial histories that continue to foreclose tribal futures in remediation design and the politics that make it difficult for institutions to speak their own fictions aloud. And it should address the exposures that are not recognized as failures at all—the ones that long ago became reasonable.

Conclusion

HERE, IN THE PLUTONIUM

It's dusk when we arrive at the B reactor. In the forty minutes it took to drive from downtown Richland to Hanford's northwest corner, the sky has traded burnt orange for indigo. I have never been on-site at night before, and at first I am absorbed by what I cannot see: the wide arc of the Columbia River, the gentle slope of Gable Mountain, the tumbleweeds performing their jaunty, insouciant ballet. All that remains is B, its cubist form glowing like a beacon in the soft desert darkness.

We are here on this September evening in 2019 to attend an oratorio at the reactor's core. About one hundred of us spill from buses chartered for the event, some pausing to stare open-mouthed at the facility, others walking confidently toward its light. I'm in that first category, frozen and staring outside of the bus, startled when someone accidentally brushes my arm. Though I have been here many times and could describe B's geometry with my eyes closed, I have never grown accustomed to this place. Instead, it is my discomfort that's familiar, a humming anxiety that begins on cue like a musical score.

Inside, I know the air will be stale and the walls will be the seafoam green of Winterfresh gum. I know there will be old safety posters from the Manhattan Project reminding employees not to discuss their work:

"Protection for all, *Don't talk*. Silence Means Security."¹ I know I will enter the cavernous room at the center of the reactor, and its three-story graphite core will take my breath away. I know I will stand in that space, heart pounding, and feel the multimillennial weight of what it has made.

We sit shoulder to shoulder for the oratorio, close enough to experience my neighbor's dry cough and then to smell the lozenge she pops in her mouth. It is an intimacy I will recall with surprising fondness six months later when drafting this chapter during the COVID-19 pandemic lockdown. In less than a year, the moment will feel ancient, an artifact from a previous era. Casual proximity with strangers will have faded like a dream.

Before the singing, there are introductions and thank-yous and a brief history of the reactor, which we learn went critical exactly seventy-five years ago yesterday. A uniformed ranger tells us that B is now part of the new Manhattan Project National Historical Park, a multisited effort to "preserve and interpret" the spaces of the bomb.²

"This is not your traditional national park," she says, a touch of humor in her voice. "We don't have majestic mountains or babbling brooks. But what we do have is amazing sites, stories, and legacies of the pursuit to build the first atomic weapon during World War II. This is the first full-scale production nuclear reactor in the world. Now that's a mouthful, but when we start thinking about this reactor, this tangible outcome of the Manhattan Project, what does this place mean?"

She pauses for a moment, letting the question settle like dust, then offers several more. "I've heard it referred to as the cathedral of science . . . but it can be more than science, more than a pretty amazing industrial building. This is where humankind figured out how to harness the power of the atom that fundamentally changed the world. Does this place strike dread at the destructive power of nuclear weapons? Does it strike hope, inspire hope for what nuclear medicine can bring—perhaps a cure for cancer? There are many, many perspectives, sides, and stories of even a single place like the B reactor. And those stories are spread out across the United States."

Next, Nancy Welliver, the librettist who wrote the lyrics we are about to hear, tells us she wants to talk about mythology. "A lot of times, when people say, 'it's a myth,' what they mean is, 'it's a lie that a lot of people believe.' I think that word is being misused. . . . To me, a myth is a metaphor to talk about something that's too big and too complicated to understand any other way."

Welliver worked as an environmental scientist at Hanford for almost thirty years. Part of her job, she explains, was to investigate 650 acres of radioactive landfill and to itemize the 450,000 cubic meters of waste within. Because records were often incomplete or nonexistent before 1990, she sought out retired workers who could provide insight into the burial grounds' contents. As they dug into their memories, however, the workers uncovered more than histories of waste; they also shared details about daily life on the job, including dreams they had during that time. Welliver cataloged these lived and dreamed details alongside the waste. Soon, they began to feel inseparable.

"If you work here long enough, you hear a lot of wild and crazy stories," she says, and we the audience respond with knowing laughter. "I mean you do," she nods, laughing a little herself. "I've heard my share. Some of the things you are going to hear tonight are the wild and crazy stories and some are [the workers'] dreams. I don't tell you which are which because a lot of the things that happen here are very dream-like and sometimes it's hard to tell if you are hearing a dream or you are hearing reality." That ambiguity is productive, she implies, it invites us to think about Hanford differently.

Welliver is particularly interested in the workers' dreams because she sees their mythological potential. To her, they represent a kind of "collective unconscious," a means for engaging Hanford's unreal realities. Indeed, she argues that their seeming impossibility is what makes them useful, allowing us to inhabit the contradictions of nuclear life. She hopes that trespassing the normative boundaries of image and time through song will "have a healing effect."

To that end, she asks that tonight we enter the dream-like space of Pluto's temple, which in Greece is called a Plutonium. Pluto (better known as Hades) is the god of the underworld, the god of secrets, the god of riches, garbage, and death. "And when you think about those things," Welliver says, "they are very much like Hanford: the secrets, the riches, the death, the garbage. Everything that's buried underground."

In ancient times, Plutonia were geologically active spaces, often associated with caves or fissures that emitted toxic gases. Usually, it was herdsmen who discovered them, noting changes in vegetation or unexplained animal corpses, then reporting the phenomena to a local priest. If they were deemed evidence of Pluto's breath, a temple was erected around the vapors, marking the "gates to the underworld."[3] The devout sought healing there and, if they survived its toxicity, received powerful dreams from below. "I want to put people in this kind of Plutonium, this mythological space," Welliver tells us, "before we hear this beautiful music."

The oratorio begins without words, just a sustained, resonant *ooooo* as the singers take their places between orchestra and core. The sound is palpable, the air thick and vibrating with voice. For the second time this evening, I forget to breathe. Later, I will read that composer Reginald Unterseher intended this first, simple melody to move us "into a shared story state." Its "repeating notes" and "descending melodic fragment[s]" are meant to be transportive, carrying us down "into our dream temple . . . and the physical ground."[4]

Here, in the Plutonium, the dream-songs are lovely and disorienting, catastrophic and strange. Some are playful, others aching with grief. One is a drum that pounds faster and louder with each line, at once heartbeat and countdown:

> I come home from my job
> with a terrible headache and fever.
> I go to my old bedroom
> in my parents' house in Richland
> to rest.
> I dream I am ten years old again.
> I can feel an atomic bomb, black, in the
> shape of the Little Boy bomb. I can feel it,
> it is literally inside my head
> about to explode.[5]

The bomb is sensible in the Plutonium—its shape and color, its fierce potential energy—because here, body and technology are one and the same. The little boy *is* Little Boy, "head about to explode," both destroyer and destroyed.[6]

Another song is sticky, trailing residues of fear and sound throughout the room:

> When I worked at Hanford, years ago,
> I was often in contaminated buildings
> Dressed in whites.
> I was anxious.
> I dreamed I had gotten contamination
> Inside me.
> Especially my lungs and I'd try and try
> to cough but I couldn't,
> the contamination stayed inside me
> and I was so frightened.
> Or I would dream that I had contamination
> On my feet, didn't know it
> And I'd track it all over my home.
> Dreams like that, they were tough
> I still have them sometimes.[7]

It is the "kind of nightmare where just when you need to run, you are suddenly in slow motion," Unterseher writes, his composition filling the space with cello, baritone, and nerves.[8] Like contamination, the dreamsong lingers—"I still have them sometimes"—neither substance fully going away. I feel its music acutely, my own body become sensory device: reactive, reaction, reactor.

When I interview Welliver several months later, she tells me that these songs are a corporeal form of reckoning. "I think dreams arise from the body," she says. "They arise from the bodies of these people that put their bodies into Hanford so, to me, the dreams *are* Hanford embodied. You know, people go in there and they put these PPEs on and they go into this really strange environment that's sometimes dangerous.[9] And then they have these dreams that come up out of their bodies. That creates a mythology for this place. A mythology that is unique to Hanford."[10]

As she speaks, I imagine the dreams forming in tissue, sweat, and breath. I imagine them upwelling in the Plutonium through song. And I think about what it means for contaminated land to be fleshy and relational, for the body to *be* Hanford and its waste. Among the abstract statistical models that usually narrate production and cleanup, the notion comes as a relief. Yet such embodiment feels suffocating at the same time, another uneven burden of the bomb.

In the oratorio's final act, for example, dreamers participate in remediation by eating and drinking nuclear waste:

> I am inside
> one of the Hanford high level waste tanks.
> I don't know how I got there.
> I feel trapped.
> I am frightened and lonely.
> It is dark.
> I cannot find the way out on my own.
> But I know the only way out is this: if everybody
> who has ever seen Hanford
> would eat
> just one
> spoonful of the waste
> The tanks would disappear.
> And the Hanford Site restored to beauty.
> To eat the waste; hard to bear the thought.
> But I am alone and it starts with me.[11]

Welliver tells me that many of the dreams she collected revolve around this theme. "There was a lot of stuff about eating plutonium." And though she cannot say with certainty, she thinks this may reflect "the workers' sense that they are making the waste disappear during the cleanup effort"—a transpositional labor that "reverberates in their bodies."[12]

Consuming waste also evokes the physical and emotional strains of the job. "A lot of people work out their anxieties through dreams and one of the things that I know about Hanford from having worked there for so long is that it is a really stressful place to work. Super stressful." It isn't just the daily risks of the job, she explains, but also the profound implications of the place itself.

When she began inventorying Hanford's radioactive landfills, the scale alone was a source of practical and existential dread. "I had to dig through all of these old documents that talked about stuff that had happened. You know, there was an explosion here and they buried the equipment, or they tore out this facility and they put all of the equipment in the ground. And I just did not know that there was all of this pretty horrifying stuff out there.... [I]t kind of blew my mind. I would look at some of the liquid disposal areas and I would see how much plutonium was in there and I would cry. I would think, 'I can't believe this. This is crazy.'"

Another anxious reality, of course, is that cleanup does not actually make waste disappear (that magic trick is only possible in dreams). Remediation doesn't erase; it manages and contains, in part, by imagining the body as reasonably exposed. I think about the dreamer in the high-level waste tank realizing that she is "the only way out"—that escape means becoming a containment device herself. If everyone "would eat just one spoonful of the waste," she sings, then Hanford's beauty would be restored. The melody is slow, her voice traveling upward on violin strings; hesitant at first, then resolved. "To eat the waste; hard to bear the thought. But I am alone and it starts with me."

That evening in the Plutonium has stayed with me, its songs like traces on my fingers as I write. There is the ode to the mutant rabbit someone once spotted on site ("It is bizarre, like nothing I have seen before") and the chorus of atomic soldiers crouching beneath a mushroom cloud ("The colors were beautiful. I hate to say that").[13] There are the songs for the Wanapum who have lived with this land since time immemorial and the ones for the workers who spend their days among its buildings and burial grounds. And there are the discordant refrains for the nuclear project itself: sounds of absurdity and hope, horror and awe.

> O my love, we sculpted a guitar from the sweet
> sugar sand
> In the Hanford desert.
> It played a haunting tune, then swelled
> and burst in the sun.
> The small boy cried, I know not out of sorrow,
> or from wonder.[14]

But the piece I have found most difficult to shake is called "Putting the Genie Back in the Bottle." In it, the dreamer notices smoke wafting up from his basement and hurries below ground to investigate. He finds "two strange men, laughing and careless, pouring liquid from one of the high level waste tanks into a bottle" and blanches when they offer him a swig of the sludge. The smoke obscures the men's faces as they pour waste into more bottles, these ones "labeled with the Mr. Clean genie." Unterseher describes the melody as a "menacing ostinato in 5/8" time, a mutation of the Mr. Clean jingle I vaguely remember from childhood.[15] Its final line is surprised, a single note that echoes around the room: "Mr. Clean winks at me."

I didn't feel the song's impact at first. It was short and bizarre; less moving than others in the oratorio. I barely mentioned it in the notes I scribbled on the bus ride back to Richland that night. However, as weeks have turned to months and now years, the song has grown brighter, like a Polaroid slowly coming into focus. I keep picturing the rebottled tank waste and Mr. Clean's confident gaze. I keep returning to that final wink.

Indeed, it was the first image that came to mind in December 2021, when DOE reinterpreted the statutory definition of high-level waste (HLW), making it possible to reimagine large volumes of nuclear material at Hanford.[16] Within this new frame, DOE may choose to reclassify HLW as low-level waste (LLW), a shift that could significantly alter disposal requirements. For decades, the plan has been to turn most of Hanford's tank waste into glass, a more stable form designed to withstand weathering and time.[17] If HLW is reclassified as LLW, however, DOE will have the option to immobilize this waste in a cement-like substance called grout (a quicker, cheaper, and arguably less-effective remedy).[18]

The agency has spent years making the case for reclassification. It has held public meetings across Oregon and Washington, with familiar PowerPoint slides containing clip art humans and probability curves. It has presented arguments to the HAB about grout's remedial potential too, passing a small, puck-shaped sample of the material around the room. One scientist used Will Smith's career to explain the grouting process, referencing a timeline of album covers and movie posters. First, you mix water with raw cement powder (*Rock the House*), he said, then that fresh grout (*The Fresh Prince of Bel-Air*) hardens (*Bad Boys*) and sets (*Big Willie Style*), becoming a "solid mass" (*Men in Black*) and, ultimately, a mature "final product that slowly ages" (*Suicide Squad*).[19] These discussions have felt oddly dreamlike in their own right: the grout's impossible promise in my palm, Will Smith as waste form.[20]

State agencies, tribal nations, and community organizations have resisted reclassification, citing incomplete data and long-term threats to human health and the environment. "DOE has made efforts to guarantee,

Grout sample at Hanford Advisory Board meeting, 2017. Photo by author.

via a suite of models, that receptors will be safe for the lifetime of those wastes," Oregon's Nuclear Safety Division staff wrote in 2018. "We have argued repeatedly that these models leave out key features and processes observable in the real world, supported by decades of data and evidence from DOE's own reports."[21] Reclassifying waste would not only be "contrary to law," Hanford Challenge argued that same year, but a "technically indefensible" sleight of hand.[22] Or, as Columbia Riverkeeper's conservation director Dan Serres put it, "If you're in a marathon at mile three, you can't just stop the race and say, 'I won' by moving the finish line. Sure, the race is over much faster if you move the endpoint, and it costs a lot less" but the waste doesn't go away.[23]

As often happens, I have been frustrated by the terms of the debate—discomfited that, in opposing reclassification, I find myself reiterating current structures that I *also* find inadequate. I want more than existing logics of nuclear impact, more than Cold War remedies for contaminated life. I want so much more than this.

Perhaps that is why I keep returning to Mr. Clean's wink. His song serves as a reminder that remediation is a social product: a way of seeing waste that has been made and can therefore be unmade. This flexibility feels hazardous on the one hand because existing policies have been hard won and continue to perform important cleanup work. Over the years, they have required that DOE move waste away from the river, cocoon reactors, demolish contaminated buildings, pump and treat contaminated groundwater, and build better disposal facilities. And as reclassification has shown, denaturalizing current policies could allow the agency to implement less-protective remedies like grout (completing projects by moving the finish line).

However, in the heat of such debates, it can be easy to forget that cleanup is *already* an imperfect construction. Hanford's finish lines are already fraught with inequities and erasures, its fences already filled with holes. Indeed, the very notion of an end point for eternal waste is a wink in and of itself, a normative bottle for impact and time. It imagines an "after" by bounding what the future can be.

I began this book with a simple, three-line story of cleanup. One that pretends a shared sense of purpose and narrative completion. One that is essential and also not enough. I end here, in the Plutonium, because its dream-songs offer another way into the telling and make me feel hopeful about Hanford for the first time in years. Not because they are lighthearted or comforting (they are neither of those things), but because they tell the impossible stories that I long to hear. They sing the both/and qualities of waste, of contamination, of risk—the real/surreal, waking/dreaming, possible/impossible politics of the bomb.

Critically, the oratorio embraces surreality while remaining grounded in the material. It presents the dream world as physical: its myths and metaphors originating in the body, at once uncanny *and* mundane.[24] This form of surreality does not make the bomb unthinkable, but instead highlights the powerful fictions that have dreamed it into being. And in the process, it makes space for other narrative possibilities.

Interdisciplinary scholar Ruha Benjamin insists that nonnormative relations require nonnormative stories. "We should acknowledge that most people are forced to live inside someone else's imagination," she says. "This means that for those of us who want to construct a different social reality, one grounded in justice and joy, we can't only critique the world as it is, we have to work on building the world as it should be."[25] The oratorio's dreamsongs embody the kind of "novel fictions" that Benjamin describes: "not falsehoods, but refashionings" that "reimagine and rework all that is taken for granted about the current structure" of social life.[26]

Such refashionings are not simply about newness, historian M. Murphy writes, but "also about what has to end, what has to come apart, to make less violent worlds."[27] The novel fictions I want cleanup to write are ones that make it impossible to ignore tribal rights, impossible to withhold compensation and care from injured workers, impossible to pretend that No Trespassing signs are enough protection for multimillennial waste. I want a future inhabited by more than statistical people. I want a cleanup that unmakes the bomb's reasonable harm.

Acknowledgments

My gratitude for the people and places that made this book possible far exceeds the few pages that follow. To those named and unnamed in these brief acknowledgments, please know that there is so much more I would like to say.

Thank you first and foremost to the people who perform the essential and imperfect labor of Hanford cleanup on a daily basis. I am particularly grateful to Jeff Burright, Tom Carpenter, Shelley Cimon, Dirk Dunning, Dale Engstrom, Laura Hanses, Rebecca Holland, Emmitt Jackson, Russell Jim, Emy Laija, Susan Leckband, Todd Martin, Ken Niles, John Price, Tom Rogers, Dan Serres, Tom Sicilia, and Ginger Wireman for modeling advocacy across generations. Your friendship and mentorship—the countless meals, meetings, and conversations we have shared—have meant the world to me. My deepest thanks to Liz Mattson, who has the unique gift of creating community wherever she goes, and without whom this book simply would not be. Thank you, thank you, friend of my heart, for everything.

My research and writing have benefited from personal and intellectual support at multiple academic institutions. At University of Oregon, Peter Walker and Lise Nelson provided key feedback in this project's infancy, while Paul Blanton, Kate and Jon Day, Matthew Derrick, Kevin and Lu Emerson, Chris Holman, Demian Hommel, Emily Knowles, Pollyanna Lind, Mike Prusila, Hunter Shobe, and Chuck Snyder made Eugene home. At University of California Berkeley, Jake Kosek, Alastair Isles, Nathan Sayre, Kim TallBear, and Richard Walker grounded my analysis with steady reminders to think about place and power relationally.

I am especially grateful to Jake for encouraging me to write creatively, to ask difficult questions, and to take impossibility seriously. His warmth and wisdom have buoyed me more times than I can count.

Endless thanks to my fellow graduate students at UC Berkeley for many years of friendship and conversation: Javier Arbona, Jenny Baca, Teo Ballve, Tripti Bhattacharya, Rachel Brahinsky, Liz Carlisle, Alasdair Cohen, Alicia Cowart, Jennifer Devine, Lindsey Dillon, Zoe Friedman-Cohen, Anthony Fontes, Ilaria Giglioli, Jenny Greenburg, Katy Guimond, Camilla Hawthorne, Leigh Johnson, Julie Klinger, Freyja Knapp, Sarah Knuth, Nicole List, Greta Marchesi, Andrea Marston, Jeff Martin, Nathan McClintock, Maywa Montenegro, Meredith Palmer, Shaina Potts, Annie Shattuck, John Stehlin, Adam Romero, Alex Tarr, Sapna Thottathil, Albert Velazquez, Mary Whelan, Max Woodworth, and Jerry Zee. Thank you especially to Erin Collins, whose care and insight have made this book (and my life) infinitely better.

I am fortunate to work with an amazing group of staff, students, and faculty at University of Washington Bothell. Many thanks to my colleagues in the School of Interdisciplinary Arts and Sciences for your creativity, warm counsel, and collaborative approach. Thank you in particular to my undergraduate research assistant Elizabeth Zigon for her energy and organizational skills, and to the Geography Writing Group—Kessie Alexandre, Christian Anderson, Carrie Freshour, Ben Gardner, Maryam Griffin, Jin-Kyu Jung, Santiago Lopez, Melanie Malone, Margaret Redsteer, Adam Romero, and Cleo Wölfle Hazard—for reading many iterations of this book over the years. Your thoughtful questions and moral support have been invaluable.

My sincere thanks to the generous institutions and organizations that have helped make this project possible. The Mellon/ACLS Dissertation Completion Fellowship, the Society of Women Geographers National Fellowship, and the Oregon State University Special Collections and Archives Research Center Resident Scholar Award provided essential resources for my graduate work. The Society of Scholars Fellowship and First Book Fellowship at the Walter Chapin Simpson Center for the Humanities as well as a writing residency at the Vermont Studio Center gave me the entwined gifts of time and community as this book was coming into being. Special thanks to Kathy Woodward, Rachael Arteaga, and the Simpson Center staff for nurturing this vibrant intellectual space and for being tireless advocates of public scholarship.

Thank you to Hriman and Padma McGilloway and Gail and Joe Wenaweser, for providing a home away from home during my early research trips, and to Carla Battles, Christy Hofsess, April Hulvershorn, and Michelle Melisko, for supporting my physical and mental well-being. My heartfelt thanks to Keturah Bouchard for lovingly reading an early draft of this book in its most vulnerable state, and to Jane Anne Staw for helping me cultivate a nourishing writing practice.

ACKNOWLEDGMENTS

I am deeply grateful to Holly Barker, David Bolingbroke, Dean Chahim, Alissa Cordner, Angela Day, Pedro de la Torre III, Jake Hamblin, Gabrielle Hecht, Lochlann Jain, Shiloh Krupar, Valerie Kuletz, Max Liboiron, Irene Lusztig, Joseph Masco, Danny Noonan, Trisha Pritikin, Linda Richards, Jesse Oak Taylor, and Jim Thomas for mentorship and conversations that have enriched my work. I also want to thank Jessica Berkeley, Keturah Bouchard, Adrian Chevraux-FitzHugh, Renee Chevraux and Richard FitzHugh, Stephanie Cram, Johanna Crane, Andrea Fitanides, Mollie Fleming, Shari Franjevic, Kathleen Grady, Wilder Harper, William Hartmann, Heidi Hill, Emily Knowles, the Lindsley family, Jaime Macadangdang, Gita Matlock, Laura McGinnis, Marissa Mika, Brian and Lisa Powers, Jenny Rinzler, Frank Romero, Julie Shayne, Glory Waschevski, and Berkeley Williamson for their love, friendship, and support.

It has been a privilege to work with Chad Attenborough, Andrea Butler, Naja Pulliam Collins, Stacy Eisenstark, Chloe Layman, Teresa Iafolla, and Julie Van Pelt, who expertly guided this book through the editorial and publication process at University of California Press. Thank you to Sharon Langworthy for superb copyediting, Jon Dertien for facilitating key production details, Liz Mattson and Miya Burke for careful proofreading, and Shyama Helin for the beautiful painting on this book's cover. So too, profound thanks to Rebecca Lave, who has been a trusted mentor and advocate throughout the life of this project, including as coeditor of the Critical Environments series, and to Jeff Burright, Erin Collins, Shiloh Krupar, Joseph Masco, Liz Mattson, and Ken Niles whose thoughtful comments and suggestions have improved this book (all errors remain my own).

The gratitude I feel for Jeffrey Cram, Maya Durie, Heidi Cram, Anarion Woods, and M Woods exceeds the boundaries of written expression. To my incredible family, for your love and friendship, compassion and encouragement, bravery and beauty, I thank you with all of my heart.

Finally, I am grateful every single day that I get to spend my life with Adam Romero. Thank you for your unconditional love and bright mind, your ridiculously good cooking, and your boundless care for growing things. Words cannot capture my affection.

Notes

INTRODUCTION: ON TELLING IMPOSSIBLE STORIES

1. The Hanford Nuclear Reservation is a former weapons complex in southeastern Washington State. Built in 1943 as part of the Manhattan Project, it was the first facility in the world to make plutonium on an industrial scale. It fueled the Trinity bomb in Alamogordo, New Mexico and then Fat Man a few weeks later in Nagasaki, Japan. After World War II it continued to make plutonium for more than forty years, reprocessing thousands of tons of uranium fuel. By the time it closed its final reactor in 1987, the site had manufactured 67.4 metric tons of weapons-grade plutonium—more than 60 percent of the material in the U.S. nuclear arsenal. See Gerber, Michele Stenehjem. *On the Home Front: The Cold War Legacy of the Hanford Nuclear Site.* Lincoln: University of Nebraska Press, 2002; Gephart, R. E. *Hanford: A Conversation about Nuclear Waste and Cleanup.* Columbus, OH: Battelle Press, 2003; Gephart, Roy. "A Short History of Waste Management at the Hanford Site." *Physics and Chemistry of the Earth* 35, no. 6 (2010): 298–306; and Niles, Ken. "The Hanford Cleanup: What's Taking So Long?" *Bulletin of the Atomic Scientists* 70 (2014): 37–48.

2. Indeed, according to Washington State's Office of the Attorney General, "Hanford holds more high-level radioactive waste than all other U.S. states combined." Washington State Attorney General. "Hanford." https://www.atg.wa.gov/hanford.

3. Gerber, Michele Stenehjem. *Legend and Legacy: Fifty Years of Defense Production at the Hanford Site*. Richland, WA: Westinghouse Hanford, 1992.

4. It can be difficult to pin down exact, real-time estimates for Hanford's waste. These numbers came from the *Hanford Natural Resource Damage Assessment Injury Assessment Plan*. Richland, WA: Hanford Natural Resource Trustees, 2013, 2–9; Oregon Department of Energy. "Hanford Groundwater." https://www.oregon.gov/energy/safety-resiliency/Pages/Hanford-Groundwater.aspx; and Niles, "The Hanford Cleanup."

5. Johnson, A. R., and M. J. Elsen. "An Integrated Biological Control System at Hanford." Presentation to the Hanford Advisory Board, Richland, WA, April 15, 2010; and Johnson, A. R., J. G. Cardiff, R. F. Giddings, et al. "An Integrated Biological Control System at Hanford." Presentation to the Waste Management Symposia, Phoenix, AZ, March 8, 2010.

6. Wald, Matthew. "Even Rabbit Droppings Count in Nuclear Cleanup." *New York Times*. October 14, 2009; and Scheck, Justin. "Bunnies Are in Deep Doo-Doo When They 'Go Nuclear' at Hanford." *Wall Street Journal*. December 23, 2010.

7. U.S. Department of Energy. "Hanford Site Groundwater Monitoring Report for 2020." DOE/RL-2020-60, Rev. 0. Richland, WA, 2021; U.S. Department of Health. "Hanford and Public Health." 2018. https://doh.wa.gov/community-and-environment/radiation/radiation-topics/hanford-and-public-health; and Pacific Northwest National Laboratory. "How Is Groundwater Cleanup Progressing?" https://phoenix.pnnl.gov/phoenix/apps/remediation/index.html.

8. Aaker, Grant, and Josh Wallaert. *Arid Lands*. Sidelong Films (production company), Bullfrog Films (publisher), 2007; Parshley, Lois. "Cold War, Hot Mess." *Virginia Quarterly Review* 97, no. 3 (2021): 46–71; and Schneider, Keith. "Washington Nuclear Plant Poses Risk for Indians." *New York Times*. September 3, 1990.

9. The Hanford Advisory Board (HAB) is a multistakeholder body that writes policy advice about cleanup for the U.S. Department of Energy, the U.S. Environmental Protection Agency, and the Washington State Department of Ecology. Board stakeholders include three area tribes (the Confederated Tribes and Bands of the Yakama Nation, the Confederated Tribes of the Umatilla Indian Reservation, and the Nez Perce Tribe); local government and business interests; the Hanford workforce; local and regional citizen, environmental, and public interest groups; local and regional universities; the State of Oregon; and members of the public at large. U.S. Department of Energy. "Hanford Advisory Board." https://www.hanford.gov/page.cfm/hab.

10. U.S. Department of Energy. "About Hanford Cleanup." https://www.hanford.gov/page.cfm/AboutHanfordCleanup.

11. *Oxford English Dictionary*, 3rd ed. Oxford: Oxford University Press, 2009. S.v. "Remedy."

12. Plutonium, for example, has a half-life of approximately 24,000 years. This means it will take about 240,000 years for it to reach background radiation levels.

13. My language is intentional here: one speck of plutonium in the lungs *can* cause cancer. It may or may not. This book examines the power-laden relations of such uncertainty. Centers for Disease Control. "Radioisotope Brief: Plutonium." https://www.cdc.gov/nceh/radiation/emergencies/isotopes/plutonium.htm; and U.S. Environmental Protection Agency. "Radionuclide Basics: Plutonium." https://www.epa.gov/radiation/radionuclide-basics-plutonium#plutoniumhealth.

14. Though the Department of Energy is only required to use a one-thousand-year compliance period, it nonetheless often evaluates ten-thousand-year scenarios. The Nuclear Regulatory Commission uses a ten-thousand-year scenario, as does the Environmental Protection Agency. U.S. Department of Energy. "Time of Compliance for Disposal of Low-Level Radioactive Waste." Office of Environmental Health, Safety, and Security. August 11, 2014. https://www.energy.gov/ehss/downloads/time-compliance-disposal-low-level-radioactive-waste; U.S. Nuclear Regulatory Commission, Issuing Body. *The Nuclear Waste Policy Act of 1982*. Washington, D.C.: U.S. Nuclear Regulatory Commission, 1983; and U.S. Environmental Protection Agency. "Public Health and Environmental Radiation Protection Standards for Yucca Mountain, Nevada." 40 CFR Part 197. *Federal Register* 73, no. 200 (October 15, 2008).

15. The U.S. Environmental Protection Agency defines institutional controls as "administrative and legal controls that help minimize the potential for human exposure to contamination and/or protect the integrity of the remedy." U.S. Environmental Protection Agency. "Superfund: Institutional Controls." https://www.epa.gov/superfund/superfund-institutional-controls.

16. Such activities might include building a home, eating food from a home garden, pumping groundwater for an on-site well, and so forth.

17. U.S. Environmental Protection Agency. "Residual Risk: Report to Congress." EPA-453/R-99-001. Research Triangle Park, NC: Office of Air Quality Planning and Standards. March 1999.

18. U.S. Environmental Protection Agency. "Conducting a Human Health Risk Assessment." n.d. https://www.epa.gov/risk/conducting-human-health-risk-assessment#tab-5.

19. My grandfather worked at Hanford for a short time during the Cold War as a heavy-equipment operator (building roads). Most of his stories from that time centered around dust.

20. As Supreme Court Justice Stephen Breyer once put it, cancer is "the engine that drives much of health risk regulation." Breyer, Stephen. *Breaking the Vicious Circle: Toward Effective Risk Regulation*. Cambridge, MA: Harvard University Press, 1993, 6.

21. Jain, S. Lochlann. *Malignant: How Cancer Becomes Us*. Berkeley: University of California Press, 2013, 186.

22. Proctor, Robert, and Londa L. Schiebinger. *Agnotology: The Making and Unmaking of Ignorance*. Stanford, CA: Stanford University Press, 2008.

23. For a helpful discussion of "problem closure," see Guthman, Julie. *Weighing In: Obesity, Food Justice, and the Limits of Capitalism*. Berkeley: University of California Press, 2011, 15–16.

24. Author interview. Olympia, WA. February 21, 2012.

25. As part of my research for this book, I interviewed more than one hundred people involved in Hanford's cleanup: current and former Hanford workers, site managers, city officials, community activists, academics, research scientists, federal and state regulators, staff at environmental restoration and waste management offices for three local area tribes, members of the Hanford Advisory Board, and others. Though these interviews were unscripted, I often relied on a core set of questions, including "Will Hanford ever be cleaned up?" Most people answered with a version of "no" or "it depends what you mean by 'clean.'" Here are a few representative examples:

- "Progress is possible, but to get all of the way done? No."
- "No. Never. There will always be contamination here, it will never be cleaned up. Nobody likes to say this out loud, but it is contaminated forever. . . .There's no place else to take it, no place to put it. They will guard it, hopefully, and guard it well. And I'm not depressed by that because . . . there will still be people who care about the environment to do the best we can. But I don't think we can ever get all of the contamination out of there. I don't think we can. I don't think it's possible."
- "No, I think you'd have to take back the second law of thermodynamics, OK? (laughing) Because the reality is, we've dispersed hundreds of thousands, millions of gallons of effluent contaminant into the soils in this geology and the geology itself forbids us from retrieving it. It's not going to happen. And so, then the question is, where do we draw the perimeter? What do we try to defend from this site and how do we approach achieving what is achievable? . . . [Hanford is] the most radioactively contaminated site on this continent so the question then is how do we go about remediation and how do we do it as quickly as possible so we can defy as much as possible that very reality of physics."
- "There's no quick fix, there really isn't. There's no quick and easy fix to any of this stuff. And it's just flat not going to happen. As much as I hate to say it, I mean, Hanford's still going to be around probably trying to pump out waste, when my boys have kids and their kids have kids. I hate to say it, but it's true."
- "Last week this couple asked me, 'how long is [the waste] going to be there?' and I said, 'Forever. It will be there forever.' Well, they were horrified by this idea. They said, 'But you know, where does it go then? It's just kind of swept up? You're just going to go put it somewhere?' Well, we've created these things and they don't go away. We can isolate materials from human beings and the environment. We will do that successfully, I think. I won't say that we'll be one hundred percent successful because I don't believe in one hundred percent, I think that's too hard of a metric. But you know, high nineties."
- "Well, one day it will be pristine again but nature needs help. I know it's going to be a while but we can't just sit back and accept this, hoping that's clean enough."
- "Depends on what you mean by cleaned up. I think it will get to a point, well, I think it *can* get to a point where the waste will be sufficiently isolated to not pose

lots of public health threats. It will still pose health threats, they will just be managed ... [but] if you define cleanup like the tribes define cleanup, absolutely not, it will never be done. It just won't. And I think I've been doing this long enough that I've come to terms with that."
- "A lot of the conversations we have are not operating in reality. We need to figure out a realistic but not bad solution. We need to be willing to talk about how impossible some of the projects are."

26. As a National Research Council report put it in 2000, "Engineered barriers have limited design lives" and "Institutional Controls will fail." National Research Council. *Long-Term Institutional Management of U.S. Department of Energy Legacy Waste Sites*. Washington, D.C.: National Academies Press, 2000, 97. Here, I'm also in conversation with a growing body of scholarship that reckons with waste as a permanent (though shifting and socially produced) form. See, for example, Liboiron, Max, Manuel Tironi, and Nerea Calvillo. "Toxic Politics: Acting in a Permanently Polluted World." *Social Studies of Science* 48, no. 3 (2018): 331–49; Balayannis, Angeliki. "Toxic Sights: The Spectacle of Hazardous Waste Removal." *Environment and Planning D: Society & Space* 38, no. 4 (2020): 772–90; Lepawsky, Josh. *Reassembling Rubbish: Worlding Electronic Waste*. Cambridge, MA: MIT Press, 2018; and Joyce, Rosemary. *The Future of Nuclear Waste: What Art and Archaeology Can Tell Us about Securing the World's Most Hazardous Material*. Oxford: Oxford University Press, 2020. Importantly, I also want to note that the "impossibility" of thinking in multimillennial time scales is an administrative constraint of the settler colonial state. Hanford's tribal communities have documented changes in this landscape for thousands of years and continue to frame cleanup and restoration efforts in intergenerational time. As one tribal elder told me when I asked how future generations will remember Hanford, "We've been here since the beginning of time, we plan to be here until the end of time. We'll tell them." Author interview. Leavenworth, WA. June 28, 2012. See also State and Tribal Government Working Group. *Closure for the Seventh Generation: A Report from the State and Tribal Government Working Group's Long-Term Stewardship Committee*. Denver, CO. National Conference of State Legislatures, 2017.

27. Feminist philosopher Judith Butler writes that life's precarity "imposes an obligation" upon those who seek to care for it: "We have to ask about the conditions under which it becomes possible to apprehend a life or set of lives as precarious, and those that make it less possible, or indeed impossible." *Unmaking the Bomb* considers how such definitional frames assert their authority even as they can't quite contain the recognition they convey. It asks how the frame's power and its vulnerability are coproduced. Butler, Judith. *Frames of War: When Is Life Grievable?* London: Verso, 2010, 1.

28. As historian Jennifer L. Morgan writes, "The impossibility of recovery is inextricable from the moral imperative to attempt it." Morgan, Jennifer L.

Laboring Women: Reproduction and Gender in New World Slavery. Philadelphia: University of Pennsylvania Press, 2004, 199.

29. For a helpful discussion about working "within and against" institutions and ideas, see Harney, Stefano, and Fred Moten. *The Undercommons: Fugitive Planning and Black Study*. Wivenhoe, UK: Minor Compositions, 2013.

30. Murphy, M. *Sick Building Syndrome and the Problem of Uncertainty: Environmental Politics, Technoscience, and Women Workers*. Durham, NC: Duke University Press, 2006; Hecht, Gabrielle. "The Work of Invisibility: Radiation Hazards and Occupational Health in South African Uranium Production." *International Labor and Working-Class History* 81, no. 81 (2012): 94–113; and Balayannis, "Toxic Sights."

31. As Hepler-Smith writes, such molecular bureaucracy "came to structure environmental law and politics through, first, the efforts of 1960s U.S. policymakers to render toxic hazards subject to government control through computer-based information coordination and, second, a vision of chemical holism within the nascent US Environmental Protection Agency and the Toxic Substances Control Act, which sought to accommodate the global environment to rational administration by aggregating diverse toxic hazards and reframing them as abstract chemical substances" (535). Hepler-Smith, Evan. "Molecular Bureaucracy: Toxicological Information and Environmental Protection." *Environmental History* 24, no. 3 (2019): 534–60.

32. Malone, Melanie. "Seeking Justice, Eating Toxics: Overlooked Contaminants in Urban Community Gardens." *Agriculture and Human Values* 39, no. 1 (2022): 165–84; Kandic, Slavica, Susanne J. Tepe, Ewan W. Blanch, Shamali De Silva, Hannah G. Mikkonen, and Suzie M. Reichman. "Quantifying Factors Related to Urban Metal Contamination in Vegetable Garden Soils of the West and North of Melbourne, Australia." *Environmental Pollution* 251 (2019): 193–202; and Gómez, Hernán F., Dominic A. Borgialli, Mahesh Sharman, Keneil K. Shah, Anthony J. Scolpino, James M. Oleske, and John D. Bogden. "Blood Lead Levels of Children in Flint, Michigan: 2006–2016." *Journal of Pediatrics* 197 (2018): 158–64.

33. Shapiro, Nicholas, Nasser Zakariya, and Jody Roberts. "A Wary Alliance: From Enumerating the Environment to Inviting Apprehension." *Engaging Science, Technology, and Society* 3 (2017): 575–602, 581. See also Dillon, Lindsey. "Race, Waste, and Space: Brownfield Redevelopment and Environmental Justice at the Hunters Point Shipyard." *Antipode* 46, no. 5 (2014): 1205–21; and Puar, Jasbir K. *The Right to Maim: Debility, Capacity, Disability*. Durham, NC: Duke University Press, 2017.

34. Agard-Jones, Vanessa. "Bodies in the System." *Small Axe* 17, no. 3 (2013): 182–92; and Hoover, Elizabeth. *The River Is in Us: Fighting Toxics in a Mohawk Community*. Minneapolis: University of Minnesota Press, 2017.

35. Murphy, M. "What Can't a Body Do?" *Catalyst: Feminism, Theory, Technoscience* 3, no. 1 (2017): 1–15, 3–4.

36. Edwards, Paul N. *The Closed World: Computers and the Politics of Discourse in Cold War America.* Cambridge, MA: MIT Press, 1997, 14.

37. Masco, Joseph. "'Survival Is Your Business': Engineering Ruins and Affect in Nuclear America." *Cultural Anthropology* 23, no. 2 (2008): 361–98; and Orr, Jackie. *Panic Diaries: A Genealogy of Panic Disorder.* Durham, NC: Duke University Press, 2006.

38. Atomic Heritage Foundation. "Nevada Test Site Downwinders." July 31, 2018. https://www.atomicheritage.org/history/nevada-test-site-downwinders.

39. Nixon, Rob. *Slow Violence and the Environmentalism of the Poor.* Cambridge, MA: Harvard University Press, 2011.

40. Hurley, Jessica. *Infrastructures of Apocalypse: American Literature and the Nuclear Complex.* Minneapolis: University of Minnesota Press, 2020, 8.

41. Ibid., 9.

42. Ibid.

43. U.S. Department of Energy. *Closing the Circle on the Splitting of the Atom: The Environmental Legacy of Nuclear Weapons Production in the United States and What the Department of Energy Is Doing about It.* Washington, D.C.: U.S. Department of Energy, Office of Environmental Management; U.S. GPO, distributor, 1995.

44. Schwartz, Stephen I. *Atomic Audit: The Costs and Consequences of U.S. Nuclear Weapons since 1940.* Washington, D.C.: Brookings Institution Press, 1998.

45. Rhodes, Richard. *The Making of the Atomic Bomb.* New York: Simon & Schuster, 1986. 294.

46. Schwartz, *Atomic Audit*.

47. Alvarez, Robert. "Energy and Weapons in 2009: How Do We Assure a Sustainable, Nuclear-Free Future?" Town Hall, Seattle, WA, June 10, 2009. The Department of Energy was the successor to the Atomic Energy Commission.

48. The Office of Environmental Restoration and Waste Management is now called the Office of Environmental Management. U.S. Department of Energy. *Linking Legacies: Connecting the Cold War Nuclear Weapons Production Processes to Their Environmental Consequences.* Washington, D.C.: Office of Environmental Management, 1997, 1.

49. O'Leary was outspoken in her campaign to change the culture of nuclear secrecy in the United States. Her Openness Initiative declassified and reviewed more than three million documents about nuclear science, manufacturing, and testing. It also prompted a series of public hearings and the formation of the Advisory Committee on Human Radiation Experiments (ACHRE). Lee, Gary. "Letting the Nation in on Decades of Secrets." *Washington Post.* March 31, 1994; and Welsome, Eileen. *The Plutonium Files: America's Secret Medical Experiments in the Cold War.* New York: Dial Press, 1999. For a useful critique of such classification categories, see Masco, Joseph. "'Sensitive but Unclassified': Secrecy and the Counterterrorist State." *Public Culture* 22, no. 3 (2010): 433–63.

50. Welcome, *Plutonium Files*, 424.

51. U.S. Department of Energy, *Closing the Circle*; U.S. Department of Energy. *Estimating the Cold War Mortgage: The Baseline Environmental Management Report*. Washington, D.C.: Office of Environmental Management, 1995; and U.S. Department of Energy, *Linking Legacies*.

52. Kuletz, Valerie. *The Tainted Desert: Environmental Ruin in the American West*. New York: Routledge, 1998; and Schwartz, *Atomic Audit*.

53. U.S. Department of Energy, *Closing the Circle*, ix.

54. Grumbly's vision of time stands in stark contrast to the *Bulletin of the Atomic Scientists'* famous "Doomsday Clock." Designed to convey imminent civilizational collapse as a result of things like nuclear weapons and climate change, the Doomsday Clock ticks closer to midnight as humankind nears total ruin. See Xin, Ling. "Bulletin of the Atomic Scientists Moves Doomsday Clock 2 Minutes Closer to Midnight." *Science* January 22, 2015.

55. Hecht, Gabrielle. *Being Nuclear: Africans and the Global Uranium Trade*. Cambridge, MA: MIT Press, 2012.

56. In Richland (Hanford's hometown), residents often poke fun at such assumptions. As a retired Hanford scientist once told me with a wink, "Other people, especially the west side of the state, have an attitude that the world is ending and it's horrible and our feeling is really quite different. You know, we've worked with the materials, we're not afraid of them. Everything has its risks just like everything else does. . . . I've had six kids born here, raised here. I've got five grandchildren living here. And only one has an extra arm so that's pretty good!" Author interview. Richland, WA. August 25, 2006. For a cultural analysis of Blinky, see Schwab, Gabriele. *Radioactive Ghosts*. Minneapolis: University of Minnesota Press, 2020.

57. Brinda Sarathy and Vivien Hamilton describe a similar experience when bringing their students to visit the Stringfellow Acid Pits (designated as a Superfund site in 1983). Sarathy, Brinda, Vivien Hamilton, and Janet Farrell Brodie, eds. *Inevitably Toxic: Historical Perspectives on Contamination, Exposure, and Expertise*. Pittsburgh, PA: University of Pittsburgh Press, 2018.

58. Hecht, *Being Nuclear*, 8.

59. Ibid.

60. Masco, "Survival Is Your Business," 376.

61. National Academy of Sciences. "The Biological Effects of Atomic Radiation: Summary Reports." Washington, D.C.: National Research Council, 1956, 5.

62. Though adopted in practice by the NCRP at this time, "permissible dose" was not published by the U.S. National Bureau of Standards until 1954. *Permissible Dose from External Sources of Ionizing Radiation: Recommendations of the National Committee on Radiation Protection*. Washington, D.C.: U.S. Department of Commerce, National Bureau of Standards; U.S. GPO, 1954.

63. For an in-depth history of permissible dose and its conceptual development, see Walker, J. Samuel. *Permissible Dose: A History of Radiation Protec-*

tion in the Twentieth Century. Berkeley: University of California Press, 2000, 10–12.

64. Nash, Linda. "From Safety to Risk: The Cold War Contexts of American Environmental Policy." *Journal of Policy History* 29, no. 1 (2017): 1–33.

65. Ibid., 10.

66. Ibid.

67. Erickson, Paul, Judy L. Klein, Lorraine Daston, Rebecca Lemov, Thomas Sturm, and Michael D. Gordin. *How Reason Almost Lost Its Mind: The Strange Career of Cold War Rationality*. Chicago: University of Chicago Press, 2013.

68. Masco, Joseph. *The Future of Fallout, and Other Episodes in Radioactive World-Making*. Durham, NC: Duke University Press, 2021, 162.

69. Ibid.

70. "A response is generally warranted," the Environmental Protection Agency explains, if "cumulative excess carcinogenic risk to an individual exceeds 10^{-4} (using reasonable maximum exposure assumptions for either the current or reasonably anticipated future land use)." In other words, cleanup is recommended if, according to assumptions about present or future human activity, an individual's risk of developing cancer from exposure to the contamination in question is greater than one in ten thousand. U.S. Environmental Protection Agency. *Rules of Thumb for Superfund Remedy Selection*. EPA 540-R-97-013. OSWER 9355.0-69. Washington, D.C.: Office of Solid Waste and Emergency Response, 1997.

71. Foucault, Michel. *Security, Territory Population: Lectures at the College de France, 1977–1978*. New York: Picador, 2004; and Foucault, Michel. *Discipline and Punish: The Birth of the Prison*. New York: Vintage Books, 1979.

72. Haraway, Donna. *Simians, Cyborgs, and Women: The Reinvention of Nature*. New York: Routledge, 1991; Butler, Judith. *Bodies That Matter: On the Discursive Limits of "Sex."* New York: Routledge, 1993; and Barad, Karen. *Meeting the Universe Halfway: Quantum Physics and the Entanglement of Matter and Meaning*. Durham, NC: Duke University Press, 2007.

73. Indeed, such doses begin before birth. Recent studies have shown that umbilical cord blood contains forever chemicals like PFAS (per- and polyfluoroalkyl substances), which have been linked to cancer, birth defects, kidney disease, and other problems. Perkins, Tom. "'Forever Chemicals' Detected in All Umbilical Cord Blood in 40 studies." *Guardian*. September 23, 2022. For an excellent discussion of exposure in the broader entangled contexts of power, equity, and debility, see Puar, *Right to Maim*.

74. Fanon, Frantz. *Black Skin, White Masks*. Translated by Charles Lam Markmann. New York: Grove Press, 1967; and Butler, *Frames of War*.

75. When writing this book, I often returned to Saidiya Hartman's essay, "Venus in Two Acts." Hartman seeks to tell stories rendered speechless by the archive of Atlantic slavery, to "recuperate lives . . . [from] the account books that identified them as units of value, the invoices that claimed them as property, and the banal chronicles that stripped them of human features" (3). However, though

she longs to redress the violence that produced these numbered accounts, she finds the archive absent of the very voices she wants to recover. She could fill in such details with her imagination but does not want to ignore the fact of those archival silences. Storytelling, she finds, is both essential and not enough.

One of the many things I love about Hartman's essay is how she reckons with this paradox. She strains "against the limits of the archive to write a cultural history of the captive" while also "enacting the impossibility" of her efforts given the foundational violence that produced those same limits (11). "The task of writing the impossible (not the fanciful or the utopian but 'histories rendered unreal and fantastic')," she argues, "has as its prerequisites the embrace of likely failure and the readiness to accept the ongoing, unfinished and provisional character of this effort, particularly when the arrangements of power occlude the very object that we desire to rescue" (14). Here, recovery is not an end state or even an eventuality; it is a maddening, generative labor of "hope and defeat." It is the work "to both tell an impossible story and to amplify the impossibility of its telling" (11). Hartman, Saidiya. "Venus in Two Acts." *Small Axe: A Journal of Criticism* 12, no. 2 (2008): 1–14.

76. The word *contamination* contains multiple meanings. The first, which will no doubt sound familiar, is "the action of contaminating, or condition of being contaminated," specifically, "the presence of radioactivity where it is harmful or undesirable." The second definition is perhaps less well known: "the blending of two or more stories, plots, or the like into one." See *Oxford English Dictionary*, 3rd ed. Oxford: Oxford University Press, 2009. S.v. "Contamination."

1. TENDER

1. Roff, Heather M. "Gendering a Warbot: Gender, Sex, and the Implications for the Future of War." *International Feminist Journal of Politics* 18, no. 1 (2016): 1–18.

2. Young, Monica. "Meet Valkyrie, NASA's Space Robot." *Sky and Telescope.* May 17, 2017.

3. Institute of Electrical and Electronics Engineers. "Valkyrie." https://robots.ieee.org/robots/valkyrie/.

4. Waste Management Symposia. "Why Exhibit?" 2018. https://www.wmsym.org/exhibitors/why-exhibit/.

5. Waste Management Symposia. "Step Up to the Golf Simulator." *Insight* 44, no. 1 (March 18, 2018): 5.

6. Hanford's cleanup is run by two administrative units: Richland Operations (RL) and Office of River Protection (ORP). Price, John. "Hanford Cleanup Priorities Public Meeting." Presentation at the Hanford Site Cleanup Budget Priorities Public Meeting, Richland, WA, June 7, 2017; and Hanford Advisory Board. "HAB Consensus Advice #294. Subject: Hanford Site Budget." November 13, 2017.

7. Most of these costs were estimated to represent five to ten years of work.

8. Waste Management Symposia. "Why Attend?" 2018. https://www.wmsym.org/register/why-attend/.

9. Veolia. "Veolia Highlights World-Class Capabilities as Global Leaders in Radioactive Waste Industry Gather at WMS 2018." 2018. https://www.nuclearsolutions.veolia.com/en/news/veolia-highlights-world-class-capabilities-global-leaders-radioactive-waste-industry-gather.

10. This advertisement was for the company Jacobs CH2M.

11. Robomantis was designed and built by Motiv Robotics.

12. For a discussion of indoor air quality, sick building syndrome, and carpet in particular, see Murphy, *Sick Building Syndrome*; Grandia, Liza. "Carpet Bombings: A Drama of Chemical Injury in Three Acts." *Catalyst* 6, no. 1 (2020): 1–8; and Grandia, Liza. "Toxic Gaslighting: On the Ins and Outs of Pollution." *Engaging Science, Technology, and Society* 6 (2020): 486–513.

13. Cram, Shannon. "Escaping S-102: Waste, Illness, and the Politics of Not Knowing." *International Journal of Science in Society* 2, no. 1 (2011): 243–52.

14. As Mel Chen argues, "it is important to retain simultaneously a fine sensitivity to the vastly different intersectional sites in which toxicity involves itself in very different lived experiences (or deaths)—for instance, a broker's relation to 'toxic bonds' versus a farmworker's relation to pesticides" (279). Chen, Mel Y. "Toxic Animacies, Inanimate Affections." *GLQ* 17, nos. 2–3 (2011): 265–86.

15. Boyd, William. "Genealogies of Risk: Searching for Safety, 1930s–1970s." *Ecology Law Quarterly* 39, no. 4 (2012): 895–987; Wilson, Richard. "Risks Caused by Low Levels of Pollution." *Yale Journal of Biology and Medicine* 51 (1978): 37–51; and Wallace, L. A. "Human Exposure to Environmental Pollutants: A Decade of Experience." *Clinical and Experimental Allergy* 25 (1995): 4–9.

16. Heller, Stephen R., John M. McGuire, and William L. Budde. "Trace Organics by GC/MS [Gas Chromatography/Mass Spectrometry]." *Environmental Science & Technology* 9, no. 3 (1975): 210–13, 211.

17. Boyd, "Genealogies of Risk," 945–46.

18. Ibid., 966.

19. Hutt, Peter B. "Use of Quantitative Risk Assessment in Regulatory Decisionmaking under Federal Health and Safety Statutes." In *Risk Quantitation and Regulatory Policy*, edited by David G. Hoel, Richard A. Merrill, and Frederica P. Perera. Cold Spring Harbor, NY: Cold Spring Harbor Laboratory, 1985, 24–25, quoted in Boyd, "Genealogies of Risk," 947.

20. As William Boyd points out, this broader policy shift

> from endangerment to unreasonable risk thus went well beyond word choice, signaling an important reorientation by Congress toward notions of acceptable risk that would come to inform major legislative and regulatory efforts in the years ahead. Whereas the earlier language of endangerment trained attention to the actors and activities that were imposing hazards on the public, unreasonable risk

> suggested that the public should only be allowed to regulate the underlying activity if the associated risks were deemed to be unacceptable, translating almost seamlessly into a balancing of costs and benefits that some observers argued tilted all too easily in favor of industry. This change in language also reflected a very different posture toward uncertainty and the possibility of knowledge regarding complex and emerging environmental hazards. Unreasonable risk, and the balancing that it entailed, demanded a degree of quantification and precision that was largely absent in the earlier conceptions of endangerment. There was an assumption, in other words, that risks could be quantified and understood sufficiently in order to run them through risk-benefit analysis as a prerequisite for regulation.

Boyd, "Genealogies of Risk," 976–77.

21. Boyd, "Genealogies of Risk," citing the National Research Council. *Toxicity Testing: Strategies to Determine Needs and Priorities*. Washington, D.C: National Academies Press, 1984.

22. Nash, "From Safety to Risk," 24.

23. Ibid.

24. Samatar, Sofia. *Tender: Stories*. Easthampton, MA: Small Beer Press, 2017, 116.

25. *Encyclopedia Britannica*. S.v. "Valkyrie." November 28, 2022. https://www.britannica.com/topic/Valkyrie-Norse-mythology.

26. When thinking about care as double-edged, I have found the following analyses especially useful: Murphy, M. "Unsettling Care: Troubling Transnational Itineraries of Care in Feminist Health Practices." *Social Studies of Science* 45, no. 5 (2015): 717–37; Stevenson, Lisa. *Life Beside Itself: Imagining Care in the Canadian Arctic*. Oakland: University of California Press, 2014; Garcia, Angela. *The Pastoral Clinic: Addition and Dispossession along the Rio Grande*. Berkeley: University of California Press, 2010; Crane, Johanna Tayloe. *Scrambling for Africa: AIDS, Expertise, and the Rise of American Global Health Science*. Ithaca, NY: Cornell University Press, 2013; and Mika, Marissa. *Africanizing Oncology: Creativity, Crisis, and Cancer in Uganda*. Columbus: Ohio University Press, 2021.

27. "Curtiss-Wright EST Group." *Radwaste Solutions* 25, no. 1 (2018): 13.

28. "Endeavor Robotics." *Radwaste Solutions* 25, no. 1 (2018): 17.

29. "ThermoFisher Scientific." *Radwaste Solutions* 25, no. 1 (2018): 49.

30. Mark is a pseudonym.

31. Satariano, Adam. *The Robots Built to Clean Up Our Nuclear Mess*. New York: Bloomberg, 2017.

32. Ibid.

33. RESRAD Workshop. Environmental Science Division, Argonne National Laboratory, 2018.

34. Yu, C., A.J. Zielen, J.-J. Cheng, D.J. LePoire, E. Gnanapragasam, S. Kamboj, J. Arnish, A. Wallo III, W.A. Williams, and H. Peterson. *User's Manual for RESRAD Version 6*. Argonne, IL: Argonne National Laboratory, 2001, 1–6.

35. Ibid., 1–4.

36. Yu, Charley, Sunita Kamboj, Cheng Wang, and Jing-Jy Cheng. *Data Collection Handbook to Support Modeling Impacts of Radioactive Material in Soil and Building Structures*. Argonne, IL: Argonne National Laboratory, 2015, 2.

37. Ibid., 199.

38. Yu et al., *Data Collection Handbook*, 210–18 (fish) and 224–25 (leafy vegetables).

39. Though I couldn't find a specific description of food washing practices in the *Data Collection Handbook*, I confirmed this point with a staff member from EPA headquarters. Author interview (via phone). October 18, 2022.

40. EPA and DOE use different frameworks for calculating exposure. EPA's calculations are based on the concept of reasonable maximum exposure (RME), which it defines as "the highest exposure reasonably expected to occur at a Superfund site, and intended to estimate a conservative exposure case (i.e., well above average case) that is still within the range of possible exposures." U.S. Environmental Protection Agency. *Guidelines for Human Exposure Assessment*. EPA/100/B-19/001. Washington, D.C.: Risk Assessment Forum, 2019, 63. DOE's calculations are based on the "average member of the critical group," which it defines as "an individual receiving a dose that is representative of the more highly exposed individuals in the population." U.S. Department of Energy. "Radiation Protection of the Public and the Environment." DOE O 458.1. Office of Environment, Health, Safety, and Security. Washington, D.C., 2011, 5. DOE considers the average member of the critical group concept to be equivalent to the "Reference Person" concept outlined by the International Commission on Radiological Protection.

41. International Commission on Radiological Protection. *Report of the Task Group on Reference Man. ICRP Publication 23*. Oxford: Pergamon Press, 1975, 335.

42. Ibid., 3. Scientists tasked with updating Reference Man's data again in the mid-1990s considered creating a "World Reference Man" to represent non-Western populations as well. However, Mark Cristy (who was involved in the project) argued against this idea: "The large body of quality data upon which one bases anatomical, physiological, and biokinetics models are Western data on Western subjects. It is scientifically easier to have all of these models based on similar and consistent data than to mix and match and thus scale some of the data." Cristy, M. "Reference Man Anatomical Model." Conference presentation, Health Physics Society Summer School on Internal Dosimetry. Davis, CA. June 6, 1994. https://www.osti.gov/biblio/10186060.

43. International Commission on Radiological Protection, *Report of the Task Group on Reference Man*. Cited data are as follows: weight of the adult tongue (123), specific gravity of skin (48), weight of nails (59), blood content of the brain (215) composition and flow of nasal secretion (365), spermatozoa measurements (184), and dimensions of the heart (115).

44. Ibid., 337–38.

45. United Nations Scientific Committee on the Effects of Atomic Radiation. *Sources and Effects of Ionizing Radiation: UNSCEAR 2000 Report to the General Assembly*. New York: United Nations, 2000; International Commission on Radiological Protection. "The 2007 Recommendations of the International Commission on Radiological Protection: ICRP Publication 103." *Annals of the ICRP* 37, nos. 2–4 (2007): 1–332; Narendran, Nadia, Lidia Luzhna, and Olga Kovalchuk. "Sex Difference of Radiation Response in Occupational and Accidental Exposure." *Frontiers in Genetics* 10 (2019): 1–11; and Folkers, Cynthia. "Disproportionate Impacts of Radiation Exposure on Women, Children, and Pregnancy: Taking Back Our Narrative." *Journal of the History of Biology* 54, no. 1 (2021): 31–66.

46. Olson, Mary. "Disproportionate Impact of Radiation and Radiation Regulation." *Interdisciplinary Science Reviews* 44, no. 2 (2019): 131–39.

47. Xu, Xie George., and K. F. Eckerman. *Handbook of Anatomical Models for Radiation Dosimetry*. Boca Raton, FL: CRC Press/Taylor & Francis Group, 2010, 59.

48. Cristy, M. "Mathematical Phantoms Representing Children of Various Ages for Use in Estimates of Internal Dose." ORNL/NUREG/TM-367. Oak Ridge, TN: Oak Ridge National Laboratory. 1980, 7. Leading dosimetry scientists disagreed about the appropriate mass of reference breasts in mathematical models, leading to a heated back and forth in the journal *Health Physics*. Kramer R., and G. Drexler. "Representative Breast Size of Reference Female" *Health Physics* 40 (1981): 914; Cristy, M. "Representative Breast Size of Reference Female." *Health Physics* 43 (1982): 930–32; and Kramer, R., G. Williams, and G. Drexler. "Reply to M. Cristy." *Health Physics* 43 (1982): 932–35.

49. Stabin, M. G., et al. "Mathematical Models and Specific Absorbed Fractions of Photon Energy in the Nonpregnant Adult Female and at the End of Each Trimester of Pregnancy." ORNL/TM-12907. Oak Ridge, TN: Oak Ridge National Laboratory, 1995.

50. Researchers referred to this amalgam as 15-AF. Cristy, M., and K. F. Eckerman. "Specific Absorbed Fractions of Energy at Various Ages from Internal Photon Sources." ORNL/TM-8381. Oak Ridge, TN: Oak Ridge National Laboratory, 1987.

51. Stabin et al., "Mathematical Models and Specific Absorbed Fractions," 3.

52. Ibid. Mark Cristy also notes that female parameters are much less detailed than male parameters in Publication 23, "especially in terms of chemical composition of the body, largely for lack of data." Cristy, "Reference Man Anatomical Model."

53. Cristy, M. "Calculation of Annual Limits of Intake of Radionuclides by Workers: Significance of Breast as an Explicitly Represented Tissue." *Health Physics* 46, no. 2 (1984): 283–91.

54. Publication 130 further explains that "substantial uncertainty may be associated with the mean absorbed dose for tissues that are members of 'other tissue.'"

Paquet, F., G. Etherington, M. R. Bailey, R. W. Leggett, J. Lipsztein, W. Bolch, K. F. Eckerman, and J. D. Harrison. "ICRP Publication 130: Occupational Intakes of Radionuclides, Part 1." *Annals of the ICRP* 44, no. 2 (2015): 5–188, 132.

55. Here, I echo both Vanessa Agard-Jones and Celia Roberts in their call to think about toxicity in ways that do not essentialize normative ideas about sex, gender, and risk or reproduce "heteronormative fantasies about 'normal' bodies." Agard-Jones. "Bodies in the System," 190; and Roberts, Celia. "Drowning in a Sea of Estrogens: Sex Hormones, Sexual Reproduction and Sex." *Sexualities* 6, no. 2 (2003): 195–213. So too, I second Narendran et al. when they argue, "Special attention needs to be given to the radiation effects in transgender individuals, as the health and protection of this vulnerable group is definitely lagging behind." Narendran et al., "Sex Difference of Radiation Response in Occupational and Accidental Exposure," 9.

56. International Commission on Radiological Protection, "ICRP Publication 103," 14.

57. The ICRP recognizes this problem *even as* it describes Reference Person as a more equitable model: "The Commission has made a policy decision that there should be a single set of $_{WT}$ values that are averaged over both sexes and all ages. However, whilst adhering to this policy, the Commission fully recognizes that there are significant differences in risk between males and females (particularly for the breast) and in respect of age at exposure." International Commission on Radiological Protection, "Publication 103," 192–93. For further discussion on this point, see Hansson, Sven Ove. "Should We Protect the Most Sensitive People?" *Journal of Radiological Protection* 29, no. 2 (2009): 211–18 and Arjun Makhijani, Brice Smith, and Michael Thorne. "Science for the Vulnerable: Setting Radiation and Multiple Exposure Environmental Health Standards to Protect Those Most at Risk." Takoma Park, MD: Institute for Energy and Environmental Research, 2006. In addition to protecting the most vulnerable, it is important to address the broader social structures that distribute vulnerability unevenly. Here, I'm also thinking with Puar, *Right to Maim*; Coviello, Peter. "Apocalypse from Now On." In *Queer Frontiers: Millennial Geographies, Genders, and Generations*, edited by in Joseph Allen Boone. Madison: University of Wisconsin Press, 2000, 39–63; and Piepzna-Samarasinha, Leah Lakshmi. *The Future Is Disabled: Prophecies, Love Notes, and Mourning Songs*. Vancouver, BC: Arsenal Pulp Press, 2022.

58. International Commission on Radiological Protection, "Publication 103," 56.

59. Ibid.

60. Ibid., 91.

61. Ibid., 92. The principle of optimization reminds us that radiation protection is subjective. The ICRP *could* design a reference model that centers radiosensitive groups. It *could* imagine that Reference Person was born female and give her the BRCA mutation. It *could* recommend that she be young, recognizing

higher risk to children. Indeed, the North Carolina-based nonprofit organization Gender and Radiation Impact Project has proposed this very thing, advocating for a Reference Girl standard. https://www.genderandradiation.org/reference-girl.

62. Harris, S. G., and B. L. Harper. "Exposure Scenario for CTUIR Traditional Subsistence Lifeways." Department of Science & Engineering, Confederated Tribes of the Umatilla Indian Reservation. Pendleton, OR, 2004; and Ridolfi, Callie. "Yakama Nation Exposure Scenario for Hanford Site Risk Assessment." Yakama Nation Environmental Restoration Waste Management Program. Union Gap, WA, 2007.

63. RESRAD's default fish consumption rate, for example, is 5.4 kg/year (about 15 g/day). By contrast, tribal scenario rates are far higher at 620 g/day (CTUIR) and 452 g/day (Yakama Nation). Harris and Harper, "Exposure Scenario for CTUIR Traditional Subsistence Lifeways;" and Ridolfi, "Yakama Nation Exposure Scenario for Hanford Site Risk Assessment."

64. Hanford occupies the historic and contemporary home of the Yakama Nation, the Confederated Tribes of the Umatilla Indian Reservation (CTUIR), the Nez Perce Tribe, and the Wanapum. The Treaty of 1855 gives the Yakama Nation, Nez Perce Tribe, and CTUIR rights to hunt, fish, pasture horses and cattle, and gather traditional foods and medicines on this land in perpetuity. Although in 1943 the U.S. government restricted such activities "temporarily" for the war effort, the tribes argue that current remediation efforts must restore full access and respect their legal rights to use Hanford lands safely. The Wanapum did not sign a treaty with the U.S. government but remain deeply invested in Hanford's cleanup, where the Columbia River has been their home since time immemorial. For further discussion, see Wanapum Heritage Center. "Repository." https://wanapum.org/about/repository/; Yakama Nation. "Yakama Nation Treaty of 1855." https://www.yakama.com/about/treaty/; Confederated Tribes of the Umatilla Indian Reservation. "Treaty of 1855." https://www.ctuir.org/departments/office-of-legal-counsel/codes-statutes-laws/treaty-of-1855/; and Nez Perce Tribe. "History." https://nezperce.org/about/history/.

65. LaDuke, Winona. "Uranium Mining, Native Resistance, and the Greener Path." *Orion* 28, no. 1 (2009): 22; and Masco, Joseph. *The Nuclear Borderlands: The Manhattan Project in Post–Cold War New Mexico*. Princeton, NJ: Princeton University Press, 2006.

66. U.S. Environmental Protection Agency. *Staff Paper: Risk Assessment Principles and Practices*. Washington, D.C.: U.S. Environmental Protection Agency, Office of the Science Advisor, 2004, 22.

67. Ibid., 101.

68. Ibid., 22. A good example of this is fish consumption. While the EPA calculated a high-end (99th percentile) rate for the "general public" of 142.4 g/day, the Columbia River Intertribal Fish Commission (CRITFC) found a high-end

(99th percentile) rate of 389 g/day for Columbia basin tribes with treaty rights to fish from the river. Selecting a "general public" receptor would change the "normal distribution" of exposure one could receive from eating Columbia River fish. U.S. Environmental Protection Agency, *Staff Paper: Risk Assessment Principles & Practices*; Columbia River Inter-Tribal Fish Commission. *A Fish Consumption Survey of the Umatilla, Nez Perce, Yakama, and Warm Springs Tribes of the Columbia River Basin*. Technical Report 94-3. Portland, OR, 1994; and Bridgen, Pamela. "Protecting Native Americans through the Risk Assessment Process: A Commentary on 'An Examination of U.S. EPA Risk Assessment Principles and Practices.'" *Integrated Environmental Assessment and Management* 1, no. 1 (2005): 83–85.

69. It is likely that this staffer was citing a 2008 agency memo that argued the Yakama Nation's sweat lodge parameters "appear[ed] implausible." Stifelman, Marc. "Comments on the Yakama Nation Exposure Scenario for Hanford Risk Assessment." U.S. Environmental Protection Agency, Office of Environmental Assessment. January 3, 2008.

70. Author interview. March 6, 2012. Richland, WA. EPA toxicologist Marc Stifelman made a similar argument about tribal scenario inhalation rates in 2003. Stifelman, Marc. "Letter to the Editor." *Risk Analysis* 23, no. 5 (2003): 859–60. Authors of the CTUIR's exposure scenario responded to Stifelman's letter at length, refuting its use of U.S. national averages to assess risk in tribal communities: "Tribal populations are not represented by a 'high end tail' of a national melting pot of ethnicities, but discrete lifestyles protected by Treaties and/or federal Trusteeship obligations." Harper, Barbara, Brian Flett, Stuart Harris, Corn Abeyta, and Fred Kirschner. "Response to Letter to the Editor." *Risk Analysis* 23, no. 5 (2003): 862.

71. Author interview. April 27, 2013. Union Gap, WA.

72. Indeed, as Max Liboiron argues, pollution itself is a form of colonialism. Liboiron, Max. *Pollution Is Colonialism*. Durham, NC: Duke University Press, 2021. See also Shadaan, Reena, and M. Murphy. "EDC's as Industrial Chemicals and Settler Colonial Structures." *Catalyst: Feminism, Theory, Technoscience* 6, no. 1 (2020): 1–36; and Gochfeld, Michael, and Joanna Burger. "Disproportionate Exposures in Environmental Justice and Other Populations: The Importance of Outliers." *American Journal of Public Health* 101, no. S1 (2011): S53–S63.

73. As Dr. Russell Jim, program manager for the Yakama Nation's Environmental Restoration and Waste Management office said in 2003:

> The Columbia River is the lifeline of the Pacific Northwest. It has been such since the beginning of time. And now for instance you have a study by the Environmental Protection Agency that says the indigenous people have 1 chance in 50 of getting cancer from the chemicals if we continue to eat the fish from the Columbia, especially around the Hanford area as we have in the past. It makes you wonder: equity, fairness... Under the law if there is one cancer in ten thousand, something must be done. But after this release, we asked the EPA what they are going to do about this. And they asked, "Well what do you want us to do?" Well, it's obvious: we would like

to have it corrected. They said, "We don't have any money." If this happened in the suburbs of New York or Cincinnati, it would have been cleaned up yesterday.... And I think there should be funds provided by the government to investigate all of this. Instead, there are piecemeal efforts and no real concerted effort to involve the Yakama Nation on a true government-to-government basis on the situation. We are a sovereign nation. We out-rank the state of Washington in sovereignty, and yet we are treated like third-class people. And all this is related to your basic question: "What does the land mean to us?" All of this is tied together to our sovereignty, our government, our culture, our religion—all tied to the foods and medicines, our language, our way of life.

Jim, Russell. Interview by Cynthia Kelly, Tom Zannes, and Thomas E. Marceau. Atomic Heritage Foundation, "Voices of the Manhattan Project." Hanford, WA. https://ahf.nuclearmuseum.org/voices/oral-histories/russell-jims-interview/.

The EPA is currently evaluating its risk assessment methodology to address disproportionate impacts to tribal communities across the United States. For initial findings and plans for future work, see U.S. Environmental Protection Agency. "Consumption by Tribes of Plants and Animals Not Accounted for in EPA Superfund Risk Assessment Methodology." Office of Superfund Remediation and Technology Innovation (OSRTI). n.d. https://clu-in.org/conf/tio/plantcbyt_060320/. This link also includes research about unaccounted dietary, lifestyle, and ceremonial exposures. For a broader analysis of roadblocks and reforms within environmental justice bureaucracies at the EPA, see Harrison, Jill. *From the Inside Out: The Fight for Environmental Justice within Government Agencies.* Cambridge, MA: MIT Press, 2019.

74. Samatar, *Tender*, 116, 123.

75. Ibid., 123.

2. ANATOMY OF A PHANTOM

1. Breitenstein, B. D., et al. "The U.S. Transuranium Registry Report on the 241Am content of a Whole Body." *Health Physics* 49, no. 4 (1985): 559–661.

2. *Mosby's Medical Dictionary.* 7th ed. St. Louis: Mosby/Elsevier, 2006.

3. Lynch, Timothy P. "In Vivo Radiobioassay and Research Facility." *Health Physics* 100, no. 2 (2011): 35–40. Other facilities are located at the Argonne National Laboratory, Brookhaven National Laboratory, Carlsbad Environmental Monitoring and Research Center, Idaho National Laboratory, Lawrence Livermore National Laboratory, Los Alamos National Laboratory, Oak Ridge National Laboratory, Sandia National Laboratory, and the Savannah River Site.

4. For additional photos of and information about the IVRRF, see Lynch, T. P. "In Vivo Monitoring Program Manual." HNF-55649. Richland, WA: U.S. Department of Energy, 2014.

5. Lynch, "In Vivo Radiobioassay and Research Facility."

6. John is a pseudonym.

7. McClellan, R. O., and C. R. Watson. "Radiation Dosimetry of Cs-137 in Sheep Evaluated with Thermoluminescent Dosimeters." In *Hanford Biology Research Annual Report for 1964*. Pacific Northwest Laboratory (U.S. Department of Energy), 1965; Watson, C. R., and R. O. McClellan. "In Vivo Thermoluminescence Dosimetry of Gamma Rays from Ingested Cs-137 in Sheep." In *Luminescence Dosimetry*, edited by Frank H. Attix, 393-401; and U.S. Atomic Energy Commission, Division of Technical Information, National Bureau of Standards, U.S. Department of Commerce, Springfield, VA, 1967. For broader discussion about the Experimental Animal Farm as a site of social formation, see Bolman, Brad. "Pig Mentations: Race and Face in Radiobiology." *Isis* 112, no. 4 (2021): 694-716.

8. Lynch, Timothy P. "In Vivo Monitoring Program Manual." HNF-55649. Mission Support Alliance (U.S. Department of Energy). Richland, WA, 2014.

9. U.S. Department of Energy. "Phantom Library." Office of Environment, Health, Safety, and Security. https://www.energy.gov/ehss/phantom-library.

10. Lynch, "In Vivo Monitoring Program Manual," 24.

11. White, D. R. "Tissue Substitutes in Experimental Radiation Physics." *Medical Physics* 5, no. 6 (1978): 467-79.

12. Taylor, Frieda Yvonne. "History of the Lawrence Livermore National Laboratory Torso Phantom." ProQuest Dissertations Publishing, 1997, 5.

13. Ibid., iv.

14. Lynch, "In Vivo Monitoring Program Manual," 2.18.

15. Ibid., 6.

16. Hegenbart, L., Y. H. Na, J. Y. Zhang, M. Urban, and X. George Xu. "A Monte Carlo Study of Lung Counting Efficiency for Female Workers of Different Breast Sizes Using Deformable Phantoms." *Physics in Medicine & Biology* 53, no. 19 (2008): 5527-38.

17. Lombardo, Pasquale Alessandro, Anne Laure Lebacq, and Filip Vanhavere. "Creation of Female Computational Phantoms for Calibration of Lung Counters." *Radiation Protection Dosimetry* 170, nos. 1-4 (2016): 369-72, 369.

18. Vedantham, S., and A. Karellas. "SU-E-I-61: Phantom Design for Phase Contrast Breast Imaging." *Medical Physics* 39, no. 6 (2012): 3638-39; and Kiarashi, Nooshin, Adam C. Nolte, Gregory M. Sturgeon, William P. Segars, Sujata V. Ghate, Loren W. Nolte, Ehsan Samei, and Joseph Y. Lo. "Development of Realistic Physical Breast Phantoms Matched to Virtual Breast Phantoms Based on Human Subject Data." *Medical Physics* 42, no. 7 (2015): 4116-26.

19. Farah, Jad, David Broggio, and Didier Franck. "Creation and Use of Adjustable 3D Phantoms: Applications for the Lung Monitoring of Female Workers." *Health Physics* 99, no. 5 (2010): 649-61.

20. A 2003 report from the International Atomic Energy Agency described a similar technique in calibrating lung counters at Canadian facilities: "Overlaying breast tissue remains a problem, it is currently simulated with two flexible containers of a liquid tissue substitute which are temporarily fixed to the

phantom for the duration of the count." International Atomic Energy Agency. "Intercalibration of In Vivo Counting Systems Using an Asian Phantom." IAEA-TECDOC-1334. Radiation Monitoring and Protection Section. Vienna, Austria, 2003, 81.

21. The lab has maintained its accreditation status since February 1998, when the DOELAP Radiobioassay program was established. Lynch, "In Vivo Monitoring Program Manual," 33.

22. According to DOE Standard 1112-98, "Phantoms used for the [DOELAP] measurement categories are a bottle manikin absorption (BOMAB) phantom for whole body counting, calibration lung sets for the Lawrence Livermore National Laboratory (LLNL) torso phantom for lung counting, and the ANSI N44.3 thyroid neck phantom for thyroid counting." U.S. Department of Energy. "The Department of Energy Laboratory Accreditation Program for Radiobioassay." DOE-STD-1112-98. Washington, D.C., 1998, 12.

23. Farah et al., "Creation and Use of Adjustable 3D Phantoms," 649. See also Manohari, M. "Simulation of In-Vivo Monitors and VOXEL Phantoms for Establishing Calibration Factors." Ph.D. dissertation, Homi Bhabha National Institute, Mumbai, India, 2014; and Hegenbart, "Monte Carlo Study of Lung Counting Efficiency for Female Workers."

24. Hegenbart et al., "Monte Carlo Study of Lung Counting Efficiency for Female Workers," 5527.

25. Kramer, G. H., B. M. Hauck, and S. A. Allen. "Comparison of the LLNL and JAERI Torso Phantoms Using Ge Detectors and Phoswich Detectors." *Health Physics* 74, no. 5 (1998): 594–601.

26. Moore, Lisa Jean, and Adele E. Clarke. "The Traffic in Cyberanatomies: Sex/Gender/Sexualities in Local and Global Formations." *Body & Society* 7, no. 1 (2001): 57–96; and Jordanova, L. J. *Sexual Visions: Images of Gender in Science and Medicine between the Eighteenth and Twentieth Centuries*. Madison: University of Wisconsin Press, 1989.

27. Cristy, "Mathematical Phantoms Representing Children of Various Ages."

28. Such practices also reflected industry imaginations of the body as a collection of atomized parts. Breasts were not integral, but supplementary—one of many calculative units to be removed or appended at will. Here, I'm reminded of both Catherine Waldby's and Lisa Cartwright's analyses of the Visible Human Project. As Cartwright puts it, there is a paradox in "seeking to create a universal archive through which to represent and to know human biology, while rendering their respective body models with a level of specificity that may ultimately confound goals such as the establishment of a norm." Cartwright, Lisa. "A Cultural Anatomy of the Visible Human Project." In *The Visible Woman: Imaging Technologies, Gender, and Science*, edited by Paula A. Treichler, Lisa Cartwright, and Constance Penley, 21-43. New York: New York University Press, 1998, 39; and Waldby, Catherine. *The Visible Human Project: Informatic Bodies and Posthuman Medicine*. New York: Routledge, 2000.

29. Radiology Support Devices, Inc. "The Fission-Product Phantom." https://rsdphantoms.com/product/the-fission-product-phantom/.
30. Ibid.
31. Heinzerling, Lisa. Session IV: "Does the Law Tend to Favor Identified over Statistical People?" Harvard University, 7th Annual Program in Ethics and Health Conference. May 1, 2012. Or, as she put it in 2015, "The life of environmental law is the statistical life. It is barely an exaggeration to say there are no identified lives in environmental law. In fact, as it relates to human health, environmental law is arguably defined by reference to the protection of statistical, not identified, lives." Heinzerling, Lisa. "Statistical Lives in Environmental Law." In *Identified versus Statistical Lives: An Interdisciplinary Perspective*, edited by Glenn Cohen, Normal Daniels, and Nir Eyal, 174–81. Oxford: Oxford University Press, 2015.
32. Heinzerling, Lisa. "The Rights of Statistical People." *Harvard Environmental Law Review* 24, no. 1 (2000): 189–207, 189.
33. Ibid.
34. Snicket, Lemony. *Horseradish: Bitter Truths You Can't Avoid*. New York: HarperCollins, 2007.
35. *Oxford English Dictionary*. Oxford: Oxford University Press, 2000. S.v. "Phantom."
36. Nelson, I. C., K. R. Heid, P. A. Fuqua, and T. D. Mahony. "Plutonium in Autopsy Tissue Samples." *Health Physics* 22, no. 6 (1972): 925–30.
37. U.S. Transuranium and Uranium Registries. "Hanford Autopsy Study" https://ustur.wsu.edu/history/hanford-study/, citing C. E. Newton, K. R. Heid, H. V. Larson, and I. C. Nelson. "Tissue Sampling for Plutonium through an Autopsy Program." Richland, WA: Battelle Memorial Institute, Pacific Northwest Laboratory, 1966.
38. Ibid.
39. U.S. Transuranium and Uranium Registries. "National Human Radiobiology Tissue Repository." https://ustur.wsu.edu/nhrtr/.
40. This is how I have always felt about B horror films whose subjects undergo some kind of radioactive transformation. These monsters inspire terror not only because they enact widespread violence, but because their mutant forms are so extravagantly unreal. Godzilla does more than destroy large swaths of Tokyo; he threatens the boundaries of life itself. Kunimoto, Namiko. *The Stakes of Exposure: Anxious Bodies in Postwar Japanese Art*. Minneapolis: University of Minnesota Press, 2017.
41. Baker, Peter, and Choe Sang-Hun. "Trump Threatens 'Fire and Fury' against North Korea if It Endangers U.S." *New York Times*. August 8, 2017.
42. These tensions, falling as they did on the anniversary of the Hiroshima and Nagasaki bombings, felt exceptionally weighty with history.
43. Alex is a pseudonym.
44. Here, I'm thinking with Jake Kosek's "Aggregate Modernities" and the honeybee-as-algorithm. He asks: "What are the material practices by which bees

are made into data and data into apiary aggregates? What contortions and transformations does the honeybee endure to become the kind of aggregated data compatible with algorithmic processing? What are the broader political consequences of these ways of knowing and calculating life through algorithms? And, crucially, how does the very process of data collection work to constitute the authority of the experts it produces?" Kosek, Jake. "Aggregate Modernities: A Critical Natural History of Contemporary Algorithms." In *Other Geographies: The Influence of Michael Watts*, edited by Sharad Chari, Susanne Friedberg, Vinay Gidwani, Jesse Ribot, and Wendy Wolford. Newark, NJ: Wiley-Blackwell, 2017, 67.

45. In his 1917 lecture "Science as a Vocation," German sociologist Max Weber famously argued that "increasing intellectualization and rationalization" in modern science has led to the belief that "one can, in principle, master all things by calculation. This means that the world is disenchanted." Weber, Max. *From Max Weber: Essays in Sociology*. London: Routledge, 1952, 139. Building on this idea, anthropologist Nancy Scheper-Hughes writes: "Medical, forensic, and biotechnical sciences require rational and rationalized bodies—bodies that can be broken down into fragments, disarticulated, de-personalized, and rendered anonymous. Science needs bodies that are "disenchanted," in the Weberian sense, rendered into plain "things" and into reusable, recyclable raw materials. But bodies—whole and in parts—do not acquire these properties so easily." Scheper-Hughes, Nancy. "Dissection." In *A Companion to the Anthropology of the Body and Embodiment*, by Frances Mascia-Lees. Oxford: Wiley-Blackwell, 2011, 184.

46. U.S. Transuranium and Uranium Registries. "Prospective Collaborators." https://ustur.wsu.edu/collaborators/.

47. Ibid.

48. Pu is the atomic symbol for plutonium. Voelz, G. L., J. N. P. Lawrence, and E. R. Johnson. "Fifty Years of Plutonium Exposure to the Manhattan Project Plutonium Workers: An Update." *Health Physics* 73, no. 4 (1997): 611–19; and McInroy, J. F., R. L. Kathren, R. E. Toohey, M. J. Swint, and B. D. Breitenstein. "Postmortem Tissue Contents of 241Am in a Person with a Massive Acute Exposure." *Health Physics* 69, no. 3 (1995): 318–23.

49. Avtandilashvili, Maia, Stacey L. McComish, George Tabatadze, and Sergei Y. Tolmachev. "USTUR Research: Land of Opportunity." Presentation at EURADOS Annual Meeting. KIT, Karlsruhe, Germany. February 27–March 2, 2017. Another USTUR presentation I read referred to the Registry Room as the "Gold Mine." James, Tony. "DOE's U.S. Transuranium and Uranium Registries (USTUR): Studying Occupational Exposures to Plutonium from Beginning to End." Special presentation to DOE/EH. December 14, 2005.

50. Masco, Joseph. *The Theater of Operations: National Security Affect from the Cold War to the War on Terror*. Durham, NC: Duke University Press, 2014, 45.

51. Here I'm also thinking about the Cold War origins of the statistical life concept, which, as Spencer Banzhaf argues, first facilitated cost-benefit analyses

of military applications and then did the same for environmental regulations based in risk philosophies. Banzhaf, H. Spencer. "Retrospectives: The Cold-War Origins of the Value of Statistical Life." *Journal of Economic Perspectives* 28, no. 4 (2014): 213–26.

52. Welsome, *Plutonium Files*.

53. These experiments were conducted under contract with the U.S. government. President Clinton later apologized for its role in nonconsensual human experimentation, saying, "When the government does wrong, we have a moral responsibility to admit it." Goodwin, Irwin. "Clinton Apologizes for Cold War's Radiation Experiments, Which Panel Says 'Created a Legacy of Distrust' in Science." *Physics Today* 48, no. 11 (1995): 70.

54. Murphy, M. *The Economization of Life*. Durham, NC: Duke University Press, 2017, 6.

55. Ibid., 7.

56. Heinzerling, "Rights of Statistical People," 190.

57. Ibid., 89.

58. Heinzerling, Lisa. "Knowing Killing and Environmental Law." *New York University Environmental Law Journal* 14, no. 3 (2006): 521–736, 526–27.

59. Ibid., 527.

60. David Roberts (physicist and former science adviser to the U.S. ambassador to Japan) uses *micromorts* and *microlives* to discuss risks associated with the Fukushima disaster. Roberts, David. "Putting Fukushima into Perspective." *Wall Street Journal*. September 12, 2013.

61. U.S. Nuclear Regulatory Commission. "Reassessment of the NRC's Dollar per Person-Rem Conversion Factor Policy." Federal Register 80, no. 172 (September 4, 2015): 53585.

62. Viscusi, W. Kip. *Fatal Tradeoffs: Public and Private Responsibilities for Risk*. Oxford: Oxford University Press, 1996, 19.

63. Heinzerling, "Rights of Statistical People," 194; and Kysar, Douglas A. *Regulating from Nowhere: Environmental Law and the Search for Objectivity*. New Haven, CT: Yale University Press, 2010, 112–14.

64. Heinzerling, "Knowing Killing," 531–32.

65. For more information about the UPPU Club, including links to long-term studies about members' health as well as an oral history with one of the members, see "Discussions with Deb: The UPPU Club." *Cold War Patriots* (blog). July 16, 2021. https://coldwarpatriots.org/blog/discussions-with-deb-the-uppu-club/

66. *Oxford English Dictionary*. Oxford, England: Oxford University Press, 2000. S.v. "Remember."

67. The USTUR logo contains a triad of symbols: an atomic orbital; a simplified version of Leonardo da Vinci's Vitruvian Man; and a caduceus with twin serpents, staff, and wings. The Vitruvian Man presents the body as a mathematical

equation: "ideal proportions" that together seek perfect collective symmetry. The caduceus contains the staff of Hermes, a liminal god of boundaries in Greek mythology with the power to reanimate the dead. Together with the atomic orbital, these images narrate a distinct disciplinary politics of the body in the nuclear age. Foucault, Michel. *The Birth of the Clinic: An Archaeology of Medical Perception*. New York: Vintage Books, 1994.

68. Sistrom, Joseph, Jerry Hopper, Sydney Boehm, Gene Barry, Lydia Clarke, Michael Moore, Nancy Gates, Lee Aaker, Leith Stevens, and Paramount Pictures Corporation. *The Atomic City*. Paramount, 1992 (originally produced as a motion picture in 1952).

69. Burris Cunningham isolated the first pure sample of plutonium in 1942.

70. Stories about Case 102's body decomposing at a Seattle mortuary made national news. "Radioactive Corpse Left Decomposing." *Tennessean*. September 4, 1979; "Tainted Corpse Lay 4 Months in Mortuary." *Detroit Free Press*. September 4, 1979; and "Body of Nuclear Scientist Studied after 4-Month Wait." *Oregonian*. September 3, 1979.

71. McInroy, J. F., H. A. Boyd, B. C. Eutsler, and D. Romero. "The U.S. Transuranium Registry Report of the 241Am Content of a Whole Body: Part IV, Preparation and Analysis of the Tissues and Bones." *Health Physics* 49, no. 4 (1985): 587.

72. Breitenstein et al., "U.S. Transuranium Registry Report on the 241Am Content of a Whole Body," 633.

73. Much of the right half of 0102's body was cremated, dissolved, and radiochemically analyzed, while bones from the left side of his body were distributed among his anthropomorphic phantoms (with the exception of the skull and torso phantoms, which drew from his right side). Hickman, D. P., and N. Cohen. "Reconstruction of a Human Skull Calibration Phantom Using Bone Sections From an ^{241}Am Exposure Case." *Health Physics* 55, no. 1 (1988): 59–65; and Kephart, Gary Steven. "An Arm Phantom for In Vivo Determination of Americium-241 in Bone." Master's thesis, University of Washington, 1987.

74. The lawsuit listed "12 causes of action ranging from negligence to restricted liability" in 0102's death. Associated Press. "Scientist's Family Sues." *Daily Pilot*. March 28, 1980. The report found that melanoma rates were five times higher than expected among LLNL employees when compared with data from the surrounding counties. It did not assign causality to occupational exposures, however, instead making the case that additional research was needed. Joint Legislative Audit Committee. "The Cancer Incidence among Workers at the Lawrence Livermore Laboratory: A Synthesis of Expert Reviews of the Study." Sacramento, CA, 1980; McBride, Stewart. "New Life for Nuclear City." *New York Times*. January 18, 1981; and "Rate of Melanoma Higher at Atom Lab." *Washington Post*. April 23, 1980.

75. As Lisa Heinzerling writes, "Identified lives are not described in terms of individual risks posed to a population of people; that is part of the definition of statistical, not identified, lives. . . . The statistical nature of the lives protected by

regulation of air toxics thus allows a reframing of the problem—to a problem of risk, not life—which itself allows—culturally, politically, morally—a pretty casual brush-off of the lives at stake." Heinzerling, "Statistical Lives in Environmental Law," 177.

76. Author interview (via phone). June 21, 2018.

77. Alan's mother died of cancer a decade after his father did. She was fifty-nine.

78. Email correspondence. June 21, 2018.

79. Here, I am again indebted to Saidiya Hartman's critical analysis of writing with and against the archive. Hartman, "Venus in Two Acts."

80. Gordon, Avery. *Ghostly Matters: Haunting and the Sociological Imagination*. Minneapolis: University of Minnesota Press, 2008, 17.

81. Butler, *Frames of War*.

82. Indeed, one USTUR presentation described donating one's body to the Registries as a way to "Extend Your LIFE!!!" James, "DOE's U.S. Transuranium and Uranium Registries," 2005, slide 59.

83. As Kramer et al. noted, "An error in the [radio]activity estimate can be quite large if the commercial leg phantom is used to estimate what is contained in the USTUR leg phantom [Stu-0102] and, consequently, a real person." Kramer, Gary, et al. "Comparison of Two Leg Phantoms Containing ^{241}Am in Bone." *Health Physics* 101, no. 3 (2011): 248–58, 257.

84. Hickman, "Reconstruction of a Human Skull Calibration Phantom," 59.

85. Nogueira, P., and W. Ruhm. "Person-Specific Calibration of a Partial Body Counter Used for Individualized ^{241}Am Skull Measurements." *Journal of Radiological Protection* 40 (2020): 1362–89, 1370.

86. The right half of the skull—Stuart Gunn's—is contaminated; the left half from another man of similar size is not. Hickman and Cohen, "Reconstruction of a Human Skull Calibration Phantom"; and Nogueira and Ruhm, "Person-Specific Calibration of a Partial Body Counter."

87. The missing metatarsal and pinky toe bones in the leg phantom were discovered via a CT scan in 2013. The reason for their absence is unknown (unrecorded). Tabadatze, George, et al. "Re-evaluation of Am 241 Content in the USTUR Case 0102 Leg Phantom." *Health Physics* 104, no. 1 (2013): 9–14. The missing metacarpal and finger bones had been "previously committed to use in other research in progress at Lawrence Berkeley Laboratory" so could not be included in the arm phantom. However, the scientists making the phantom were able to make molds of his finger bones in order to create accurate artificial duplicates. Kephart, "An Arm Phantom for In Vivo Determination of Americium-241 in Bone," 26. The sternum bone is missing from the torso phantom because it was radiochemically analyzed.

88. Kephart, "Arm Phantom for In Vivo Determination of Americium-241 in Bone," 31.

89. Ibid., 31, 32.

90. Butler, *Frames of War*, 3.
91. Ibid., 3 and 6.
92. Gordon, *Ghostly Matters*, 21.
93. Ibid., 22.

3. RATIONAL MUTANTS

1. Creager, Angela. "Radiation, Cancer, and Mutation in the Atomic Age." *Historical Studies in the Natural Sciences* 45, no. 1 (2015): 14–48; and Crow, James. "Quarrelling Geneticists and a Diplomat." *Genetics* 140, no. 2 (1995): 421–26.
2. A roentgen is a unit of ionizing radiation.
3. The underlining is in the original. National Academy of Sciences. "The Biological Effects of Atomic Radiation: Summary Reports." Washington, D.C.: National Research Council, 1956, 29.
4. The committee considered ten roentgens to be the "doubling dose"—the rate at which radiation-induced mutations would be equal to spontaneous (nonradiogenic) mutations in the body. The doubling dose was often used in developing human exposure guidelines for radiation. However, according to Christopher Jolly, "Like most benchmarks in radiation protection, this standard was more or less arbitrary, based on the seemingly conservative assumption that doubling the spontaneous mutation rate would not significantly injure the genetic constitution of the population." Jolly, J. Christopher. "Thresholds of Uncertainty: Radiation and Responsibility in the Fallout Controversy." Oregon State University. ProQuest Dissertations Publishing, 2004, 293.
5. Here, I'm in conversation with Brinda Sarathy, Vivien Hamilton, and Janet Farrell Brodie, who argue that such "historical analysis is powerful precisely because it disrupts the sense that our current predicament is inevitable. Understanding how these spaces came into being can help us identify contemporary institutions as well as modes of thinking and acting that continue to allow environments of toxicity to persist." Sarathy et al., *Inevitably Toxic*, 5.
6. "Is This What It's Coming To?" *New York Times*. July 5, 1962. Oregon State University Special Collections and Archives Research Center, Ava Helen and Linus Pauling Papers, box 7.014, folder 14.11.
7. Cousins, Norman. "The Debate Is Over." *Saturday Review*. April 4, 1959. Oregon State University Special Collections and Archives Research Center, Ava Helen and Linus Pauling Papers, box 7.003, folder 3.24.
8. "The Contaminators." *Playboy*. October 1959. Oregon State University Special Collections and Archives Research Center, Ava Helen and Linus Pauling Papers, box 7.012, folder 12.6.
9. Palmer, Harvey et al. "Radioactivity Measurements in Alaskan Eskimos in 1963." *Science* 144, no. 3620 (1964): 859–60. Oregon State University Special

Collections and Archives Research Center, Ava Helen and Linus Pauling Papers, box 7.006, folder 6.3.

10 French, Norman R. 1960. "Strontium-90 in Ecuador." *Science* 131, no. 3417 (1960): 1889–1890. Oregon State University Special Collections and Archives Research Center, Ava Helen and Linus Pauling Papers, box 7.013, folder 13.10.

11. "Summary of Remarks Prepared by Dr. Willard Libby, Commissioner United States Atomic Energy Commission for Delivery before the Swiss Academy of Medical Sciences Symposium on Radioactive Fallout." Lausanne, Switzerland. March 27, 1958. Oregon State University Special Collections and Archives Research Center, Ava Helen and Linus Pauling Papers, box 7.002, folder 2.11; Kulp, Lawrence, and Arthur Schulert. "Strontium-90 in Man V." *Science* 135, no. 3516 (1962): 619–32. Oregon State University Special Collections and Archives Research Center, Ava Helen and Linus Pauling Papers, box 7.014, folder 14.10; and Kulp, Lawrence, Arthur Schulert, and Elizabeth J. Hodges. "Strontium-90 in Man III." *Science* 129, no. 3358 (1959): 1249–55. Oregon State University Special Collections and Archives Research Center, Ava Helen and Linus Pauling Papers, box 7.001, folder 11.12.

12. Masco, Joseph. "The Age of Fallout." *History of the Present* 5, no. 2 (2015): 137–68.

13. Welsome, *Plutonium Files*, 303. This quote also evokes histories of empire that have long been integral to nuclear production. Hamilton, Kevin, and Ned O'Gorman. "Seeing Experimental Imperialism in the Nuclear Pacific." *Media+Environment* 3, no. 1 (2021): 1–23; DeLoughrey, Elizabeth. "The Myth of Isolates: Ecosystem Ecologies in the Nuclear Pacific." *Cultural Geographies* 20, no. 2 (2013): 167–84; and Barker, Holly M. "Unsettling SpongeBob and the Legacies of Violence on Bikini Bottom." *Contemporary Pacific* 31, no. 2 (2019): 345–79.

14. Atomic Heritage Foundation. "Castle Bravo." National Museum of Nuclear Science and History. March 1, 2017. https://www.atomicheritage.org/history/castle-bravo.

15. Castle-Bravo was one of sixty-seven nuclear tests conducted by the United States in the Marshall Islands between 1946 and 1958. Marshallese communities are still reckoning with this violent colonial legacy. For further reading, see Barker, Holly M. *Bravo for the Marshallese: Regaining Control in a Post-nuclear, Post-colonial World.* Belmont, CA: Wadsworth, 2013; Jacobs, Robert A. *Nuclear Bodies the Global Hibakusha.* New Haven, CT: Yale University Press, 2022; *Nuclear Testing Program in the Marshall Islands: Hearing before the Committee on Energy and Natural Resources, U.S. Senate, on Effects of U.S. Nuclear Testing Program in the Marshall Islands,* 109th Cong., 1st Sess. July 19, 2005, 109–78; Raj, Ali. "In Marshall Islands, Radiation Threatens Tradition of Handing Down Stories by Song." *Los Angeles Times.* November 10, 2019; Takahashi, Tatsuya, Minouk Schoemaker, Klaus Trott, Steven Simon, Keisei Fujimori, Noriaki Nakashima, Akira Fukao, and Hiroshi Saito. "The Relationship of Thyroid Cancer

with Radiation Exposure from Nuclear Weapon Testing in the Marshall Islands." *Journal of Epidemiology* 13, no. 2 (2003): 99–107; Simon, Steven L., André Bouville, Charles E. Land, and Harold L. Beck. "Radiation Doses and Cancer Risks in the Marshall Islands Associated with Exposure to Radioactive Fallout from Bikini and Enewetak Nuclear Weapons Tests: Summary." *Health Physics* 99, no. 2 (2010): 105–23; and Bolton, Matthew Breay. "Human Rights Fallout of Nuclear Detonations: Reevaluating 'Threshold Thinking' in Assisting Victims of Nuclear Testing." *Global Policy* 13, no. 1 (2022): 76–90.

16. Ropeik, David. "How the Unlucky Lucky Dragon Birthed an Era of Nuclear Fear." *Bulletin of Atomic Scientists*. February 28, 2018; Higuchi, Toshihiro. *Political Fallout: Nuclear Weapons Testing and the Making of a Global Environmental Crisis*. Stanford, CA: Stanford University Press, 2020.

17. Ropeik, "How the Unlucky Lucky Dragon Birthed an Era of Nuclear Fear."

18. Higuchi, *Political Fallout*, 54.

19. Quoted in Hamblin, Jacob Darwin. "'A Dispassionate and Objective Effort': Negotiating the First Study on the Biological Effects of Atomic Radiation." *Journal of the History of Biology* 40, no. 1 (2007): 147–77.

20. Ropeik, "How the Unlucky Lucky Dragon Birthed an Era of Nuclear Fear."

21. Godzilla was a fictional product of nuclear testing in the South Pacific.

22. In future iterations, the name would change slightly to the Biological Effects of Ionizing Radiation (BEIR). The most recent iteration, BEIR VII, focused on health risks from low-level exposures. National Research Council. *Health Risks from Exposure to Low Levels of Ionizing Radiation: BEIR VII Phase 2*. Washington, D.C: National Academies Press, 2006.

23. Quoted in Hamblin, "Dispassionate and Objective Effort."

24. As Bronk tells it, this conversation over cocktails "initiated a chain reaction" that brought the BEAR study into being. National Academy of Sciences. "Minutes of Meeting: Study Group on Dispersal and Disposal of Radioactive Wastes." Washington, D.C., February 23, 1956, 7–8.

25. The scale and import of intergenerational mutation from ionizing radiation remains an ongoing subject of scientific study and debate, with an increasing focus on epigenetic mechanisms. See, for example, Chen, J., D. M. Sridharan, C. L. Cross, and J. M. Pluth. "Cellular DNA Effects of Radiation and Cancer Risk Assessment in Cells with Mitochondrial Defects." *Journal of Radiation Research and Applied Sciences* 15, no. 1 (2022): 89–94; Burgio, Ernesto, Prisco Piscitelli, and Lucia Migliore. "Ionizing Radiation and Human Health: Reviewing Models of Exposure and Mechanisms of Cellular Damage. An Epigenetic Perspective." *International Journal of Environmental Research and Public Health* 15, no. 9 (2018): 1971; and Dubrova, Yuri E., and Elena I. Sarapultseva. "Radiation-Induced Transgenerational Effects in Animals." *International Journal of Radiation Biology* 98, no. 6 (2022): 1047–53.

26. Quoted in Gabbert, Elisa. *The Unreality of Memory: And Other Essays*. New York: Farrar, Straus and Giroux, 2020.

27. National Academy of Sciences, "Biological Effects of Atomic Radiation," 5.
28. Ibid., 7.
29. Ibid., 19.
30. Ibid., 20.
31. Ibid., 29.
32. Ibid., 5.
33. Creager, "Radiation, Cancer, and Mutation in the Atomic Age," 36.
34. "Genetics Panel, Second Meeting. Chicago, IL, February 5 and 6, 1956," 7. CalTech, George Wells Beadle Papers, box 17, folder 17.1.
35. Ibid., 2.
36. Ibid., 21, 9.
37. Ibid., 9.
38. Walker, *Permissible Dose*.
39. U.S. National Bureau of Standards. *Permissible Dose from External Sources of Ionizing Radiation: Recommendations of the National Committee on Radiation Protection*. National Bureau of Standards Handbook 59. Washington, D.C.: U.S. Department of Commerce, National Bureau of Standards, U.S. GPO, 1954, 26–27. Though this was the first formal report that centered acceptable risk as the basis for radiation protection, the reasoning had been around much longer. Joseph Boland argues that it was implicit within the concept of the tolerance dose as well. Boland, Joseph B. "The Cold War Legacy of Regulatory Risk Analysis: The Atomic Energy Commission and Radiation Safety." University of Oregon: ProQuest Dissertations Publishing. 2002.
40. "Genetics Panel, Second Meeting. Chicago, IL, February 5 and 6, 1956," 8. CalTech, George Wells Beadle Papers, box 17, folder 17.1.
41. Ibid., 6, 14, 15, 27.
42. Ibid., 30.
43. Ibid., 31.
44. Ibid., 33. I can't help but wonder how this discussion about miscarriage might have been different if the Genetics Committee had included female scientists.
45. Ibid., 33.
46. Beatty, John. "Masking Disagreement among Experts." *Episteme* 3, nos. 1–2 (2006): 52–67, 63.
47. Ibid., 64
48. National Academy of Sciences, "Biological Effects of Atomic Radiation" 27.
49. Ibid., 26.
50. Ibid., 18.
51. Ibid., 18.
52. See Hamblin, Jacob Darwin. *Poison in the Well: Radioactive Waste in the Oceans at the Dawn of the Nuclear Age*. New Brunswick, NJ: Rutgers University Press, 2008; and Masco, "Age of Fallout."
53. Nash, "From Safety to Risk."

54. In recent decades, studies of endocrine-disrupting chemicals and nonmonotonic dose response curves have unsettled the regulatory ideal for chemical toxicity thresholds as well. Hill, Corinne E., J. P Myers, and Laura N. Vandenberg. "Nonmonotonic Dose-Response Curves Occur in Dose Ranges That Are Relevant to Regulatory Decision-Making." *Dose-Response* 16, no. 3 (2018): 1–4; and Welshons, Wade V., Kristina A. Thayer, Barbara M. Judy, Julia A. Taylor, Edward M. Curran, and Frederick S. Vom Saal. "Large Effects from Small Exposures. I. Mechanisms for Endocrine-Disrupting Chemicals with Estrogenic Activity." *Environmental Health Perspectives* 111, no. 8 (2003): 994–1006.

55. U.S. National Bureau of Standards, *Permissible Dose from External Sources of Ionizing Radiation*, 27.

56. Weinstein, B. L. Letter to Linus Pauling. October 12, 1959. Oregon State University Special Collections and Archives Research Center, Ava Helen and Linus Pauling Papers, box 7.012, folder 12.17; Cohen, David D. Letter to Linus Pauling. October 29, 1959. Oregon State University Special Collections and Archives Research Center, Ava Helen and Linus Pauling Papers, box 7.012, folder 12.17; and Schneer, Richard M. Letter to Linus Pauling. November 17, 1959. Oregon State University Special Collections and Archives Research Center, Ava Helen and Linus Pauling Papers, box 7.012, folder 12.17.

57. Shaw, Grazia Denig. Letter to Linus Pauling. May 24, 1961. Oregon State University Special Collections and Archives Research Center, Ava Helen and Linus Pauling Papers, box 7.014, folder 14.8.

58. Maupin, Ann M. Letter to Linus Pauling. December 2, 1959. Oregon State University Special Collections and Archives Research Center, Ava Helen and Linus Pauling Papers, box 7.012, folder 12.17.

59. Strontium-90 is a carcinogenic by-product of atomic testing that mimics calcium in the body. The St. Louis–based "Baby Tooth Survey," which identified increasing levels of Sr-90 in children, made this radionuclide a particular concern for parents in the 1950s and 1960s. Filler, Mary. Letter to Linus Pauling. February 3, 1960. Oregon State University Special Collections and Archives Research Center, Ava Helen and Linus Pauling Papers, box 7.013, folder 13.9.

60. Linton, Etta. Letter to Linus Pauling. August 13, 1962. Oregon State University Special Collections and Archives Research Center, Ava Helen and Linus Pauling Papers, box 7.014, folder 14.7.

61. Foresman, Mrs. Robert. Letter to Linus Pauling. September 22, 1959. Oregon State University Special Collections and Archives Research Center, Ava Helen and Linus Pauling Papers, box 7.012, folder 12.17.

62. For a helpful discussion about the gendered politics of managing contaminated food in nuclear contexts, see Kimura, Aya Hirata. *Radiation Brain Moms and Citizen Scientists: The Gender Politics of Food Contamination after Fukushima*. Durham, NC: Duke University Press, 2016.

63. U.S. Federal Civil Defense Administration. *Facts about Fallout*. Washington, D.C.: U.S. GPO, 1955.

64. Oakes, Guy. *The Imaginary War: Civil Defense and American Cold War Culture.* Oxford: Oxford University Press, 1994.

65. Masco, *Theater of Operations*, 52.

66. Peterson, Val. "Panic, the Ultimate Weapon?" *Colliers* 21 (1953): 99–105.

67. Orr, *Panic Diaries*.

68. Peterson, "Panic," 1–2.

69. As Shiloh Krupar writes, "Individuals are called to internalize risk as an individual attribute and must discipline themselves to be viable subjects in terms of risk (Rose 2007). Subsequently, risk is treated as the property of individuals rather than an economic, environmental, and social problem; embodied historical effects become the private and depoliticized property of individuals." Krupar, Shiloh R. "The Biopsic Adventures of Mammary Glam: Breast Cancer Detection and the Practice of Cancer Glamor." *Social Semiotics* 22, no. 1 (2012): 47–82, 55. She cites Rose, Nikolas S. *Politics of Life Itself: Biomedicine, Power, and Subjectivity in the Twenty-First Century.* Princeton, NJ: Princeton University Press, 2007.

70. U.S. Federal Civil Defense Administration. *Survival under Atomic Attack.* Document 130. Washington, D.C.: Executive Office of the President, National Security Resources Board, Civil Defense Office, 1951, 12.

71. Ibid., 12.

72. Ibid., 4.

73. Ibid., 5.

74. Ibid.

75. Ibid., 30.

76. Ibid., 27.

77. Ibid., 26.

78. Ibid., 27.

79. Ibid., 27.

80. Masco, *Theater of Operations*.

81. U.S. Federal Civil Defense Administration *Survival under Atomic Attack*, 31.

82. As William Boyd notes, "The report did not exactly provide the resolution the AEC sought but instead emphasized the potential harm from low doses of radiation." Boyd, "Genealogies of Risk," 920.

83. Jolly, "Thresholds of Uncertainty," 417.

84. Thompson, Dorothy. "Radioactivity and the Human Race." *Ladies Home Journal.* September 11, 1956.

85. Ibid.

86. Quoted in Goodman, Michael. "National Radiation Health Standards—A Study in Scientific Decision Making." *Atomic Energy Law Journal* 6, no. 3 (1964): 239. In his 1957 testimony before the JCAE, Taylor also said:

> In connection with the question of permissible dose standards, there are, as I have indicated, a large number of variables, and a large number of uncertainties. We do

not have, and probably will not have for a long time, any quantitative evaluation of the risks involved. What somebody has to do is evaluate the risk due to radiation against the benefits from the use of radiation, and the risk due to other things that affect our normal life. In this connection I frequently feel compelled to say that this question of radiation safety and permissible dosage standards is not a subject for which there is a clean and simple answer. The whole question of setting radiation exposure limits depends on physics and biology. It depends enormously on ethics and morality, and on an enormous amount of good judgment and good wisdom on the part of the people who are responsible for setting them. It is by no means a clean-cut quantitative physical problem.

The Nature of Radioactive Fallout and Its Effects on Man: Hearings before the Special Subcommittee on Radiation of the Joint Committee on Atomic Energy on the Nature of Radioactive Fallout and Its Effects on Man. 85th Cong., 1st Sess. 1957.

87. Goodman, "National Radiation Health Standards," 239.

88. Ibid.

89. Taylor, quoted in "Strontium May Be 60 Times Deadlier Than the AEC Says." *I.F. Stone's Weekly* 7, no. 17 (1959): 1–4.

90. Taylor, Lauriston. Letter to Strauss, November 10, 1948. Lauriston Sale Taylor Papers, box 31, file: "NCRP-1948." Quoted in Whittemore, Gilbert Franklin. "The National Committee on Radiation Protection, 1928–1960: From Professional Guidelines to Government Regulation." Harvard University. Pro Quest Dissertations Publishing, 1986, 442–43.

91. Taylor, Lauriston. Letter to Strauss, November 10, 1948. Lauriston Sale Taylor Papers, box 31, file: "NCRP-1948." Quoted in Whittemore, "National Committee on Radiation Protection," 442–43.

92. Nash, "From Safety to Risk."

93. Boland, "Cold War Legacy of Regulatory Risk Analysis," 539.

94. *Nature of Radioactive Fallout and Its Effects on Man: Hearings before the Special Subcommittee on Radiation of the Joint Committee on Atomic Energy* (testimony of Willard Libby), 1223.

95. Newell, R. R. "Genetic Injuries." *Radiology* 69, no. 1 (1957): 111–14. By way of example, Newell also writes that if radiation were to "escape out onto the sidewalk," one would simply multiply the milliroentgens per day by the average number of people who walked by. One wouldn't "fret about the possibility that a newsboy will stand there for six hours" (112).

96. Eisenbud, Merril. "The Risks." In *America Faces the Nuclear Age*, edited by Johnson E. Fairchild and David Landman. New York: Sheridan House, 1961, 105.

4. BODY BURDEN

1. To watch video footage of this meeting, see "Public Hearing on Exposed Workers at Hanford Nuclear Site—Parts I–III." https://www.youtube.com/watch?v=2zWqHzEw48M.

2. Gerber, *On the Home Front*.

3. Murphy, Kim. "Government Finally Hears a Nuclear Town's Horrors." *Los Angeles Times*. February 5, 2000.

4. The mushroom cloud logo has long been a source of both pride and controversy in Richland. Baker, Mike. "A Proud Nuclear Town Grapples with How to Remember the Bomb." *New York Times*. September 10, 2022; Associated Press. "School's Mushroom Cloud Stirs Up Controversy." *Associated Press News*. December 11, 1987; and Cary, Annette. "Japanese Exchange Student Offers Her Views on Richland High School's Mushroom Cloud Logo." *Seattle Times*. June 10, 2019.

5. Portions of this chapter (including a version of this opening scene) were published in Cram, Shannon. "Living in Dose: Nuclear Work and the Politics of Permissible Exposure." *Public Culture* 28, no. 3 (2016): 519–39. Duke University Press. All rights reserved. Reprinted by permission of the publisher. www.dukeupress.edu.

6. Flynn, Michael. "A Debt Long Overdue." *Bulletin of the Atomic Scientists* 57, no. 4 (2001): 43. The review was conducted by the National Economic Council and the report was issued on March 31, 2000. Warrick, Joby. "Panel Links Illness to Nuclear Work; Former Energy Dept. Employees May Get Compensation for Exposure." *Washington Post*. January 30, 2000. In addition to finding increased risk to atomic workers, the report concluded that there were "serious flaws in state worker compensation programs that covered contractor employees, which prevented many sick nuclear workers from being adequately compensated." Flynn, "Debt Long Overdue," 43.

7. Cary, Annette. "DOE Report Long on Horizon." *Tri-City Herald*. February 27, 2000.

8. Warrick, "Panel Links Illness to Nuclear Work."

9. Fletcher, Meg. "Proposal for New Benefits." *Business Insurance* 34, no. 18 (2000): 2. Uranium miners have been eligible for compensation through the Radiation Exposure Compensation Act since 1990.

10. Flynn, "Debt Long Overdue"; and Parascandola, Mark J. "Compensating for Cold War Cancers." *Environmental Health Perspectives* 110, no. 7 (2002): A404-7.

11. Brown, David. "Aid to Atomic Weapons Workers." *Washington Post*. July 16, 1999.

12. As Representative Sheila Jackson Lee said in a hearing revisiting the act six years later, "This act is an act that provides wholeness." *Energy Employees Occupational Illness Compensation Program; Are We Fulfilling the Promise We Made to These Cold War Veterans When We Created This Program?: Hearing before the Subcommittee on Immigration, Border Security, and Claims*. 109th Congress, 2nd Sess. May 4, 2006.

13. Ibid.

14. Jay is a pseudonym.

15. See National Park Service. "B Reactor: Manhattan Project National Historical Park." https://www.nps.gov/places/000/b-reactor.htm.

16. To read more about Hanford Challenge, visit https://www.hanfordchallenge.org.

17. Author interview. Seattle, WA. December 13, 2011. Hanford's "chilled work environment" has been the subject of multiple congressional hearings, including. *Whistleblower Retaliation at the Hanford Nuclear Site: Hearing before the Subcommittee on Financial and Contracting Oversight of the Committee on Homeland Security and Governmental Affairs, U.S. Senate*, 113th Cong., 2nd Sess. March 11, 2014.

18. Nalder, Eric. "The Plot to Get Ed Bricker—Hanford Whistle-Blower Was Tracked, Harassed, Files Show." *Seattle Times*. July 30, 1990.

19. National Institute for Occupational Safety and Health. "Review of Hanford Tank Farm Worker Safety and Health Programs." Centers for Disease Control and U.S. Department of Health and Human Services. November 28, 2016; U.S. Department of Energy. "Office of Enterprise Assessments Follow-Up Assessment of Progress on Actions Taken to Address Tank Vapor Concerns at the Hanford Site." Office of Worker Health and Safety Assessments, Office of Environment, Safety, and Health Assessments. January 2017; and U.S. Department of Energy. "Office of Enterprise Assessments Follow-Up Assessment of Progress on Actions Taken to Address Tank Vapor Concerns at the Hanford Site." Office of Worker Health and Safety Assessments, Office of Environment, Safety, and Health Assessments. February 2018.

20. Jain, S. Lochlann. *Injury: The Politics of Product Design and Safety Law in the United States*. Princeton: Princeton University Press, 2006; and Brown, Wendy. *States of Injury: Power and Freedom in Late Modernity*. Princeton, NJ: Princeton University Press, 1995.

21. Haraway, *Simians, Cyborgs, and Women*, 194.

22. Ibid.

23. *Employee Radiation Hazards and Workmen's Compensation: Hearings before the Subcommittee on Research and Development of the Joint Committee on Atomic Energy*. 86th Cong., 1st Sess. March 10–19, 1959.

24. *Radiation Protection Criteria and Standards: Their Basis and Use*; Summary-Analysis of the Hearings before the Special Subcommittee on Radiation of the Joint Committee on Atomic Energy. 86th Cong., 2nd Sess. May 24, 25, 26, 31 and June 1, 2, 3, 1960, 25.

25. Ibid., 58.

26. Ibid., 6.

27. Ibid., 21–22.

28. Alpha particles will only travel short distances (about one to two inches) and can be stopped by a thin sheet of paper or skin, while beta particles can move several feet and may require shielding made of plastic or metal. Gamma rays

and neutrons can easily penetrate the body, traveling hundreds of feet from their source. Though lead, water, or concrete can shield the body from some gamma radiation, workers also manage their exposure by quota—counting how much they have received each day and doing their best not to exceed weekly and annual allowances. U.S. Nuclear Regulatory Commission. "Radiation Basics." https://www.nrc.gov/about-nrc/radiation/health-effects/radiation-basics.html#.

29. *Radiation Protection Criteria and Standards*, 17.

30. Ibid., 40.

31. Walker, *Permissible Dose*.

32. "Radiation Safety Program Policies and Procedures." Los Angeles City College. October 27, 2020. https://www.lacitycollege.edu/Departments/Rad-Tech/documents/2020-LACC-Radiation-Safety-Program-(1).pdf.

33. Washington River Protection Solutions. "ALARA Work Planning." Document: TFC-ESHQ-RP_RWP-C-03, Rev. 0-2. November 30, 2017. https://hanfordvapors.com/wp-content/uploads/2018/08/TFC-ESHQ-RP_RWP-C-03-ALARA-Work-Planning.pdf; and U.S. Nuclear Regulatory Commission. "Occupational ALARA and Planning Controls." In *NRC Inspection Manual*. March 2, 2016. https://www.nrc.gov/docs/ML1534/ML15344A278.pdf.

34. Schieber, Caroline, and Christian Thézée. "Towards the Development of an ALARA Culture." *Proceedings of IRPA* 10 (2000): 1–4.

35. Ibid., 3.

36. U.S. Department of Energy. *DOE Handbook: Radiological Worker Training*. DOE-HDBK-1130-2008. Washington, D.C.: DOE, 2013, 39. https://www.standards.doe.gov/standards-documents/1100/1130-bhdbk-2008-cn2-2013-reaff-2013/@@images/file.

37. The "standard of care" is defined as using reasonable judgment to prevent harm to oneself or others. California Civil Code CACI 401, for example, states that "a person is negligent if he or she does something that a reasonably careful person would not do in the same situation." See Judicial Council of California. "Civil Jury Instructions." 2020 edition. California Civil Code 401. "Basic Standard of Care." https://www.justia.com/trials-litigation/docs/caci/400/401/.

38. Jose, Donald, and David Wiedis. "Preparing the Health Physicist to Testify at Deposition." *Radiological Safety Officer Magazine* 8, no. 1 (2003): 23–29. On the other hand, M. A. Boyd argues that employers often view ALARA goals as "operational limits, despite the fact that exceeding them is not considered a legal overexposure." Thus, he makes the case that the concept has been successful at reducing worker exposure overall. Boyd, M. A. "A Regulatory Perspective on Whether the System of Radiation Protection Is Fit for Purpose." *Annals of the ICRP* 41, no. 3-4 (2012): 57–63, 61.

39. Jose and Wiedis, "Preparing the Health Physicist to Testify at Deposition," 25.

40. U.S. Department of Energy, *DOE Handbook: Radiological Worker Training*, 26.

41. Ibid., 16.

42. Ibid., 28.

43. Clarke, Lee. *Mission Improbable: Using Fantasy Documents to Tame Disaster.* Chicago: University of Chicago Press, 1999.

44. Stephanie is a pseudonym. Author interview. Richland, WA. April 17, 2012.

45. Ibid.

46. Jenny is a pseudonym. Author interview. Leavenworth, WA. October 2, 2013.

47. Parr, Joy. *Sensing Changes: Technologies, Environments, and the Everyday, 1953-2003.* Vancouver: University of British Columbia Press, 2010.

48. Ibid., 68.

49. Sanger, S. L. *Working on the Bomb: An Oral History of WWII Hanford.* Portland, OR: Portland State University Press, 1995.

50. Arthur is a pseudonym. Author interview. Richland, WA. March 6, 2012.

51. Still another person (a legal advocate for Hanford whistleblowers) told me:

> There's been a deliberate shift to get younger, less experienced workers that are cheaper into the work force. No pension, not complaining as much, but they don't know what they're dealing with ... very little training and boom you're a Hanford worker. They are deliberately given skilled jobs like operators used to do, taking over a lot of that work and the operators are like, 'No you can't have them do that, they're not trained and you're putting them all at risk,' but they are losing those fights. That's how Hanford is being run now. And it scares the hell out of a lot of people who are retiring out of there saying 'I don't even want to go to work anymore. I'm worried about what's going to happen.' If someone doesn't know what they're doing, it can be dangerous. I'm not saying that's the way it is all over the place, but I think those incidents happen on a regular enough of a basis that people are concerned.

Author interview. Leavenworth, WA. October 2, 2013.

52. Szymendera, Scott D. *The Energy Employees Occupational Illness Compensation Act.* CRS Report: R46476. Washington, D.C.: Congressional Research Service, 2020.

53. Michaels, David. *Doubt Is Their Product: How Industry's Assault on Science Threatens Your Health.* Oxford: Oxford University Press, 2008, 230.

54. Krupar, Shiloh R. *Hot Spotter's Report: Military Fables of Toxic Waste.* Minneapolis: University of Minnesota Press, 2013, 186-87.

55. Silver, Ken. "The Energy Employees Occupational Illness Compensation Program Act: New Legislation to Compensate Affected Employees." *Workplace Health & Safety* 53, no. 6 (2005): 267-77, 268.

56. Viv is a pseudonym. Author interview. Richland, WA. May 9, 2012.

57. I heard the same thing from a Colorado-based advocate in 2012. Rather than providing the calculations, "NIOSH just tells us, 'Look we are the experts and we are using dose reconstructions that we think are very lenient.'" Author interview (via phone). April 5, 2012.

58. Maggie is a pseudonym. Author interview (via phone). January 4, 2013.

59. Silver, "Energy Employees Occupational Illness Compensation Program Act," 275.

60. Clara is a pseudonym. Author interview (via phone). April 25, 2012. Viv told me the same thing.

61. As Shiloh Krupar argues, "Even where records exist, it is impossible to know the precision of the monitoring equipment; past dosage monitoring instruments have proved extremely inadequate." Krupar, *Hot Spotters Report*, 188.

62. Dan is a pseudonym. Author interview (via phone). May 17, 2012.

63. "Dress out" is slang for getting suited up for radioactive work.

64. Placing the dosimetry badge on the inside of one's coveralls would make it difficult (if not impossible) to detect radiogenic exposure. "Crap up" is slang for getting contamination on something or someone.

65. Epidemiologist David Richardson put it like this: "The question is whether the [cancer risk values] are valid, and the validity is partly a question of extrapolating from a bomb blast to chronic exposures. That's probably the main source of uncertainty." He and a colleague at University of North Carolina–Chapel Hill found "substantially higher risks at low-level exposures" in Oak Ridge National Laboratory workers "than would have been predicted based on data from the atomic bomb survivor study." Parascandola, "Compensating for Cold War Cancers," A406. Other scientists found lower risks than predicted. Brooks, Antone L. *Low Dose Radiation: The History of the U.S. Department of Energy Research Program*. Pullman: Washington State University Press, 2018. The subject remains an ongoing matter of scientific debate. Lindee, M. Susan. *Suffering Made Real: American Science and the Survivors at Hiroshima*. Chicago: University of Chicago Press, 1994.

66. Krupar, Shiloh. "The Biomedicalisation of War and Military Remains: US Nuclear Worker Compensation in the 'Post–Cold War.'" *Medicine, Conflict, and Survival* 29, no. 2 (2013): 111–39, 127; and Krupar, *Hot Spotter's Report*, 172.

67. Krupar, *Hot Spotter's Report*, 174.

68. Ibid., 175. See also Proctor and Schiebinger, *Agnotology*.

69. Szymendera, *Energy Employees Occupational Illness Compensation Program Act*, 4.

70. Silver, "Energy Employees Occupational Illness Compensation Program Act," 268.

71. For a list of SEC classes, see https://www.dol.gov/agencies/owcp/energy/regs/compliance/law/SEC-Employees. This list includes Hanford, though the covered time period only extends to 1990 and a limited number of contractors. Szymendera, *Energy Employees Occupational Illness Compensation Program Act*.

72. Krupar, "Biomedicalisation of War and Military Remains," 116.

73. Pritikin, Trisha T. *The Hanford Plaintiffs: Voices from the Fight for Atomic Justice*. Lawrence: University Press of Kansas, 2020.

74. Jain, *Injury*, 24.

75. Ibid., 12, citing Scarry, Elaine. *The Body in Pain: The Making and Unmaking of the World.* Oxford: Oxford University Press, 1985.

76. Scarry, *Body in Pain*, 300.

77. Brown, *States of Injury*. In 2021 (more than two decades after EEOICPA's passage), a Washington State survey of sixteen hundred Hanford workers found that 57 percent had experienced a dangerous exposure event on the job and 32 percent had long-term exposure to vapors. Cary, Annette. "57% of Hanford Nuclear Site Workers Surveyed by Washington State Report Toxic Exposures." *Tri-City Herald.* July 7, 2021.

78. Jain, *Injury*, 3.

79. "Public Hearing on Exposed Workers at Hanford Nuclear Site" (YouTube).

80. This figure has increased dramatically since 2000. According to DOE, Hanford's cleanup is now estimated to cost between $323 billion and $677 billion (2019 estimate). U.S. Department of Energy. "2019 Hanford Lifecycle Scope, Schedule, and Cost Report." DOE/RL-2018-45. Richland, WA. January 2019.

81. "Public Hearing on Exposed Workers at Hanford Nuclear Site" (YouTube).

5. TRESPASSING

1. Washington State Department of Health. "Hanford Environmental Radiation Oversight Program: 2017 Data Summary Report." Publication 320-124. Office of Radiation Protection. Richland, WA, 2019.

2. U.S. Environmental Protection Agency, "Superfund Remedy Overview, Key Principles and Guidance." https://www.epa.gov/superfund/superfund-remedy-overview-key-principles-and-guidance.

3. U.S. Department of Energy. "Use of Risk-Based End States." DOE Policy 455.1. Office of Environmental Management, Washington, D.C., July 15, 2003, 1.

4. Ibid, 2.

5. As one former agency staff member told me in 2012, "Oh, of course, institutional controls are cheaper. Always, always, always . . . the institutional controls are essentially just keeping human beings away . . . the value that comes into play is land use. It's always land use. Well, what is the land use going to be? Is it going to be that the tribe is going to be able to return to a pristine environment? And, of course, the tribes don't think it is ever going to be pristine. . . no one anticipates that they are going to go back to what things were like in the pre-Hanford days. But they would like their people to be able to walk on the site and not be impacted." Author interview. Olympia, WA. February 21, 2012.

6. U.S. Department of Energy. 'Final Hanford Comprehensive Land-Use Plan Environmental Impact Statement." DOE/EIS-0222-F. Washington, D.C., 1999, 3–18.

7. Ibid.

8. Ibid, 1–32.

9. U.S. Department of Energy. "Supplemental Analysis of the Hanford Comprehensive Land-Use Plan Environmental Impact Statement." DOE/EIS-0222-SA-02. Richland, WA, 2015, 15, 26.

10. The exposure scenario designed by the CTUIR, for example, assumes "unrestricted access" to the Hanford site, explaining that this is "not a visiting scenario like a recreational scenario." Harris, "Exposure Scenario for CTUIR Traditional Subsistence Lifeways," 8.

11. U.S. Department of Energy, "Final Hanford Comprehensive Land-Use Plan," 6–14.

12. Ibid.

13. U.S. Department of Energy, "Supplemental Analysis," 2015, 2.

14. U.S. Department of Energy. "Hanford Site Cleanup Completion Framework." DOE/RL-2009-10, Rev. 1. Richland, WA, January 2013, 16–18.

15. U.S. Department of Energy. "Supplemental Analysis: Hanford Comprehensive Land-Use Plan Environmental Impact Statement." DOE/EIS-0222-SA-01. Richland, WA, June 2008, D-2.

16. Cleanup is, of course, a complex endeavor involving multiple, overlapping decision-making processes that reflect the different agencies, laws, and their regulatory requirements. It's not just the CLUP and only the CLUP. However, these multilayered processes and regulations often refer to the CLUP as foundational to land-use decisions that inform cleanup actions.

17. U.S. Environmental Protection Agency. "Memorandum: Land Use in the CERCLA Remedy Selection Process." Office of Solid Waste and Emergency Response. OSWER Directive No. 9355.7-04. Washington, D.C., May 25, 1995, 5.

18. Ibid., 7.

19. 40 C.F.R. § 1502.22. Cited in Fogleman, Valerie M. "Worst Case Analyses: A Continued Requirement under the National Environmental Policy Act?" *Columbia Journal of Environmental Law* 13, no. 1 (1987): 53.

20. U.S. Department of Energy, "Final Hanford Comprehensive Land-Use Plan," 5-1.

21. Ibid.

22. George, Marlene. "Review of the Remedial Investigation/Feasibility Study and Proposed Plan for the 100-BC-1, BC-2 and BC-5 Operable Units, DOE/RL-2010-96 and DOE/RL-2016-43, Draft A." January 11, 2018, 10. https://www.columbiariverkeeper.org/sites/default/files/2019-11/YN%20final%20comments%20-1-11-2018%20re%20100-B-C%20RIFS-PP-DOE-RL-2010-96%20-DOE-RL-2016-43-Draft%20A%20%283%29%20%281%29.pdf.

23. U.S. Department of Energy, "Hanford Site Cleanup Completion Framework," iii.

24. Ibid., xi.

25. Ibid., 19.

26. Ibid., 22.

27. Ibid.

28. In 2020 the Council on Environmental Quality (which implements NEPA) updated its regulatory definition of "reasonably foreseeable." Part of a broader move within the Trump administration to streamline federal environmental impact assessments, the new definition requires that agencies only evaluate effects that have "a reasonably close causal relationship to the proposed action" under review. This means that when agencies like DOE are assessing cleanup remedies through the NEPA process, they should not consider effects that are "remote in time, geographically remote, or the result of a lengthy causal chain." Council on Environmental Quality. "Update to the Regulations Implementing the Procedural Provisions of the National Environmental Policy Act." *Federal Register.* 85 FR 43304. Washington, D.C., July 16, 2020, 43344.

29. U.S. Department of Energy. "Hanford Long-Term Stewardship Program Plan." DOE/RL-2010-35, Rev. 1. Richland, WA, April 2012, 1-1. Once all cleanup activities on-site are complete, Hanford's site-specific LTS program will shift management authority to DOE's national Office of Legacy Management.

30. U.S. Environmental Protection Agency. "Comprehensive Five-Year Review Guidance." Office of Solid Waste and Emergency Response. OSWER No. 9355.7-03B-P. Washington, D.C., July 17, 2001, iv.

31. In 2012 Dr. Russell Jim, program manager for the Yakama Nation's ERWM office, told the EPA's National Remedy Review Board (NRRB), "We are all acquainted with the concept of the letter of the law, and the spirit of the law. One involves strict interpretation of words, the other, intent. This matter before us, I believe, involves both. The intent of the hazardous waste cleanup laws surely came from a desire and intent to return the land and resources to their unpolluted conditions, as nearly as possible. The term protectiveness involves the need for safety and security from harm, but only makes sense in the context of how people might reasonably live in the future." Dr. Jim's testimony takes the federal government's language seriously, using reasonableness to make the case for more cleanup, rather than less. If "protectiveness" under CERCLA is truly measured by "how people might reasonably live in the future," then the agencies need only look to the Treaty of 1855, which ensures that tribal nations may use this land in perpetuity. It's right there, he reminds the NRRB, in both the letter and the spirit of the law. Columbia Riverkeeper. "Competing Visions for the Future of Hanford." June 2018, 4. https://www.columbiariverkeeper.org/sites/default/files/2018-07/2018.6.1%20Hanford%20Vision%20Report_j_INTERACTIVE.pdf.

32. For a broader analysis of the Office of Legacy Management and Long-Term Stewardship in the context of the Rocky Flats site, see Krupar, Shiloh R. "Alien Still Life: Distilling the Toxic Logics of the Rocky Flats National Wildlife Refuge." *Environment and Planning D: Society & Space* 29, no. 2 (2011): 268–90.

33. Author interview. Richland, WA. March 6, 2012.

34. Personal communication (via email). November 11, 2022.

35. U.S. Environmental Protection Agency. Superfund Site Remediation Program: Radiation Risk Assessment Training. Phoenix, AZ, March 18, 2018.

36. U.S. Department of Energy. "Hanford Annual Site Environmental Report for Calendar Year 2020." DOE/RL-2021-15. Richland, WA, 2021, A-9.

37. Ibid., 2-39–2-40.

38. Parshley, "Cold War, Hot Mess," 54.

39. Associated Press. "'Radioactive... Do Not Eat' This Jam—Jarring Shipment to Governor, Energy Department." *Seattle Times*. August 8, 1990. According to one long-time HAB member, Buske once brought the jam to a board meeting as well.

40. Schneider, Keith. "Washington Nuclear Plant Poses Risk for Indians." *New York Times*. September 3, 1990.

41. Later, when I looked into the "great gator escape," I found conflicting stories about what happened. The best recorded information comes from oral histories with Gary Petersen (who worked in public relations and dealt with some of the associated media attention) and scientist William (Bill) Bair, who worked at Hanford's Experimental Animal Farm and conducted studies on the alligators. Neither Bair nor Petersen can remember the exact date of the incident, but both guess that it occurred in 1963. This date is confirmed by a short speech in 1971 by H. A. Kornberg, former Hanford Biology Lab director and Bair's boss at the time. Bair remembers that five alligators escaped (three from the irradiated research group and two from the control research group), while Petersen remembers that there were six. Bair says that two alligators (one irradiated and one control) were never found, while Petersen says all were found. Bair recalls that the alligators squeezed through the fence, while Petersen says the fence was blown down in a big storm. Val Tyler, who worked with Bair and was part of the alligator search party, said that the alligators dug underneath the fence. Stang, John. "Hanford's Great Gator Escape Genuine." *Tri-City Herald*. May 26, 2002; Peterson, Gary. "Interview with Gary Peterson." Hanford Oral History Project. Washington State University-Tri-Cities. Richland, WA, June 5, 2014; and Bair, William. "Interview with Bill Bair." Hanford Oral History Project. Washington State University-Tri-Cities. Richland, WA, August 14, 2013.

The detail about the fisherman and the alligator differs with each storyteller as well. While all parties agree that a local man caught one of the escapees and had it stuffed, Bair and Tyler say it was hanging in a local sporting goods shop when officials from General Electric (Hanford's general contractor at the time) confiscated it. Petersen says that a technician from the Experimental Animal Farm was walking by a Pasco taxidermy shop when he noticed a stuffed alligator through the window. "So he ran in, grabbed the alligator, and ran out." Petersen also says that when he took the public relations job at Hanford, his boss had him return the stuffed alligator to the fisherman who had caught it. He recalls successfully finding the man (who he remembers was named Aaron) and giving the alligator back. However, a 1971 Battelle Annual Report includes a transcript of a short speech by Kornberg, who is holding a stuffed alligator as he speaks. Kornberg recalls, "Eight years ago, we were using them [alligators] in our experimental program. We observed that alligators have two distinct characteristics: they hiss at managers and laboratory directors and are great escape artists. To commemorate a year made especially eventful by those characteristics, one of the alligators was stuffed, mounted, and presented to me at Biology's 1963 Christmas party. This handsome little fellow has been my constant companion ever since." Thompson, R. C. *Pacific Northwest National Laboratory Annual Report for 1971 to the USAEC Division of Biology and Medicine*. AEC Research and Development Report. BNWL-1650-PT1. Richland, WA, September 1972, 13.

According to the *Tri-City Herald*, the alligator studies began in 1961 when the first four arrived from Georgia's Okeefanokee Swamp. In the following seven months, Hanford acquired twenty-nine more from a Louisiana farm, bringing the number up to thirty-three. By the time the program was running at full capacity, it housed fifty-five alligators. Bair's experiments concerned radiosensitivity. He found that even at very high doses, no alligators died in the first thirty days following their exposure, though their health declined significantly. They experienced lethargy; partial paralysis; ulceration around the mouth; low white blood cell count; weight loss; loss of sperm count; fluid in the lungs and abdomen; abscesses in lung, liver, and stomach; and decreased bone marrow activity (all symptoms of acute radiation sickness). After thirty days, the alligators began dying very quickly—between thirty-one and sixty-three days after their exposure (relative length of life appeared relational to strength of dose). Stang, "Hanford's Great Gator Escape Genuine" and Bair, William. "The Radiosensitivity of Alligators." In *Pacific Northwest National Laboratory Annual Report for 1967 to the USAEC Division of Biology and Medicine*, edited by R. C. Thompson, Paulette Teal, and Evelyn Swezea. AEC Research and Development Report. BNWL-714. Richland, WA, May 1968.

42. Parts of this story were excerpted from Cram, Shannon. "Wild and Scenic Wasteland: Conservation Politics in the Nuclear Wilderness." *Environmental Humanities* 7, no. 1 (2016): 89–105. Duke University Press. All rights reserved. Reprinted by permission of the publisher. www.dukeupress.edu.

43. Tracy is a pseudonym. Author interview. Benton County, WA. April 17, 2012.

44. U.S. Department of Energy. "Fall 1998 200 East Area Biological Vector Contamination Report." HNF-3628. Richland, WA, March 1999.

45. Dirkes, R. L., R. W. Hanf, and T. M. Poston. *Hanford Site Environmental Report for Calendar Year 1998*. Prepared for the U.S. Department of Energy. PNNL-12088. September 1999.

46. U.S. Department of Energy, "Fall 1998 200 East Area Biological Vector Contamination Report," ix.

47. Ibid.

48. Dr. Hermann Muller, who provided the first evidence of radiation-induced mutation in 1927 (using fruit flies) and won a Nobel Prize for his studies in 1946, was frustrated by cinematic representations of radiogenic monsters like the giant ants in the film *Them!* "The popular idea that a mutation ordinarily results in a monster or a freak is a gross distortion of the facts," he wrote in 1958. This misconception, he argued, allows people to ignore the risk of smaller genetic impacts that grow increasingly harmful with time. Muller, Herman. "The Radiation Danger." *Colorado Quarterly* 6 (1958): 229–54, 232; Ashton, Linda. "Hanford's Nuke Site Produces 'Hot' Bugs." *Sacramento Bee*. October 22, 1998; "Bugs May Be Spreading Radiation at Nuke Plant." *Orlando Sentinel*. October 8, 1998; and Associated Press. "Radiation Bugging Hanford." *Spokesman Review*. October 8, 1998.

49. U.S. Department of Energy, "Fall 1998 200 East Area Biological Vector Contamination Report," F-6, F-7; and author interview. Benton County, WA. April 17, 2012.

50. Ibid., I-3.

51. Ibid., F-6.

52. Gerber, Michele Stenehjem. "All-Out Effort Stems Spreads of Contamination." *Hanford Reach*. October 19, 1998.

53. U.S. Department of Energy, "Fall 1998 200 East Area Biological Vector Contamination Report," F-20.

54. Several months later, the agency initiated a site-wide program with "new policies and procedures to unify the control of biota." Johnson et al., "Integrated Biological Control System at Hanford." DOE continues to spend millions managing contaminated flora and fauna. Wald, "Even Rabbit Droppings Count in Nuclear Cleanup."

55. Masco, Joseph. "Mutant Ecologies: Radioactive Life in Post–Cold War New Mexico." *Cultural Anthropology* 19, no. 4 (2004): 517–50.

56. National Research Council. *Long-Term Institutional Management*, 5.

57. Ibid., 98–99.

58. Ibid., 98.

59. Ibid.

60. For an excellent in-depth discussion about this dynamic, see de la Torre III, Pedro. "Unmaking Wastelands: Inheriting Waste, War, and Futures at the Hanford Site." Rensselaer Polytechnic Institute. Pro Quest Dissertations Publishing, 2017.

61. Author interview. Seattle, WA. February 13, 2012.

62. National Research Council, *Long-Term Institutional Management*, 97.

CONCLUSION: HERE, IN THE PLUTONIUM

1. An image of this poster is available on the DOE's Declassified Document Retrieval System (DDRS). Accession #N1D0023596. https://reading-room.lab works.org/Files/GetDocument.aspx?id=N1D0023596.

2. *H.R. 1208, to Establish the Manhattan Project National Historical Park in Oak Ridge, TN, Los Alamos, NM, and Hanford, WA: Legislative Hearing before the Subcommittee on Public Lands and Environmental Regulation of the Committee on Natural Resources, U.S. House of Representatives*, 113th Cong., 1st Sess., April 12, 2013.

3. See Barras, Colin. "This Roman 'gate to Hell' Killed Its Victims with a Cloud of Deadly Carbon Dioxide." *Science* February 16, 2018; and Pfanz, Hardy, Galip Yüce, Ahmet H. Gulbay, and Ali Gokgoz. "Deadly CO2 Gases in the Plutonium of Hierapolis (Denizli, Turkey)." *Archaeological and Anthropological Sciences* 11, no. 4 (2018): 1359–71.

4. "Nuclear Dreams: An Oral History of the Hanford Site." Performed by the Mid-Columbia Mastersingers. September 27, 2019. Richland, WA. Quotes and lyrics are from the Nuclear Dreams program, no page numbers provided.

5. Ibid.

6. The bomb dropped on Hiroshima, Japan.

7. "Nuclear Dreams."

8. Ibid.

9. PPE is personal protective equipment.

10. Author interview (via phone). April 27, 2020.

11. "Nuclear Dreams."

12. This description reminds me of Sofia Samatar's tender and the way her own monitoring work reverberated in her body. Samatar, *Tender: Stories*.

13. "Nuclear Dreams." Welliver took this quote from a 2019 *New York Times* documentary by Morgan Knibbe called *The Atomic Soldiers*. Knibbe, Morgan. "The Atomic Soldiers." *New York Times Ops-Docs*. Season 6. March 16, 2019. https://www.nytimes.com/2019/02/12/opinion/atomic-soldiers.html.

14. Ibid. The dreamer presented this narrative to Welliver in the form of poetry—drawing from Czeslaw Milosz's poem "Encounter." Miłosz, Czesław. *The Collected Poems, 1931–1987*. New York: Ecco Press, 1988.

15. "Nuclear Dreams."

16. U.S. Department of Energy. "Assessment of Department of Energy's Interpretation of the Definition of High-Level Radioactive Waste." *Federal Register* 86, no. 242 (2021).

17. This process is called vitrification.

18. Hanford Challenge, Natural Resources Defense Council, and Columbia Riverkeeper. "Comments on Draft Waste Incidental to Reprocessing Evaluation for Closure of Waste Management Area C at the Hanford Site." November 7,

2018. https://static1.squarespace.com/static/568adf4125981deb769d96b2/t/608
352e267faf3797471c50a/1619219171924/NRDC+et+al+Draft+WIR+Comment+
%26+Attachments+sm.pdf; and "National Academies Panel Says More Work Is
Needed on Hanford Waste Report." *Exchange Monitor*. November 9, 2022. Thank
you to Jeff Burright and Liz Mattson for helpful conversations about the reclassification process. All mistakes in describing it here are my own.

19. Asmussen, Matthew. "Grout Properties and Research Advancements." Pacific Northwest National Laboratory. PNNL-SA-14583. Presentation to the Hanford Advisory Board. Richland, WA, September 18, 2019.

20. Representatives from the Yakama Nation's Environmental Restoration
and Waste Management (ERWM) office have argued that grouting the tanks
would violate the Treaty of 1855. "We have the privileges and rights secured to
our tribal people to hunt, fish, and gather food in usual and accustomed places,"
ERWM's education and outreach specialist Alfrieda Peters said. "Cleanup needs
to occur, and to Yakama Nation standards. The Yakamas lived here before it
was Washington or before it was Oregon Territory. We still have rights there,
[and] the U.S. has an obligation to protect and preserve them." Shinn, Lora. "As
the DOE Abandons a Toxic Mess Threatening the Columbia River, the Yakama
Nation Fights Back." Natural Resources Defense Council. September 19, 2019.
https://www.nrdc.org/stories/doe-abandons-toxic-mess-threatening-columbia
-river-yakama-nation-fights-back.

21. Niles, Ken. "Comments on Draft Waste Incidental to Reprocessing Evaluation for Closure of Waste Management Area C at the Hanford Site." October 4, 2018. https://static1.squarespace.com/static/568adf4125981deb769d96b2/t
/5bc8f7a624a6949d51e08835/1539897254716/2018-10-4-ODOE-Comments-WIR
-Proposal.pdf.

22. Hanford Challenge et al., "Comments on Draft Waste Incidental to Reprocessing," 1.

23. Shinn, "As the DOE Abandons a Toxic Mess."

24. This livelier framing does not simply relegate the body to metaphor in the
ways Susan Sontag famously described. Sontag, Susan. *Illness as Metaphor*. New
York: Farrar, Straus and Giroux, 1978.

25. Benjamin, Ruha. "Race to the Future: Rethinking Innovation, Inequity,
and Imagination in Everyday Life." Keynote address at Emerson College. Boston, MA. October 17, 2019.

26. Benjamin, Ruha. "Speculative Futures: Envisioning and Creating Social
and Reproductive Justice in These Times." Keynote address at University of California at Davis. October 31, 2018.

27. Murphy, "What Can't a Body Do?," 11.

References

Aaker, Grant, and Josh Wallaert. *Arid Lands*. Sidelong Films (production company), Bullfrog Films (publisher), 2007.
Agard-Jones, Vanessa. "Bodies in the System." *Small Axe: A Journal of Criticism* 17, no. 3 (2013): 182–92.
Alaimo, Stacy. *Bodily Natures: Science, Environment, and the Material Self.* Bloomington: Indiana University Press, 2010.
Alvarez, Robert. "Energy and Weapons in 2009: How Do We Assure a Sustainable, Nuclear-Free Future?" Town Hall, Seattle, WA, June 10, 2009.
American Nuclear Society. *Radwaste Solutions* 25, no. 1 (2018): 4–108.
Asmussen, Matthew. "Grout Properties and Research Advancements." Pacific Northwest National Laboratory. PNNL-SA-14583. Presentation to the Hanford Advisory Board. Richland, WA, September 18, 2019.
Associated Press. "Scientist's Family Sues." *Daily Pilot*. March 28, 1980.
———. "School's Mushroom Cloud Stirs Up Controversy." *Associated Press News*. December 11, 1987.
———. "'Radioactive . . . Do Not Eat' This Jam—Jarring Shipment to Governor, Energy Department." *Seattle Times*. August 8, 1990.
———. "Radiation Bugging Hanford." *Spokesman Review*. October 8, 1998.
Atomic Heritage Foundation. "Castle Bravo." National Museum of Nuclear Science & History. March 1, 2017. https://www.atomicheritage.org/history/castle-bravo.

———. "Nevada Test Site Downwinders." July 31, 2018. https://www.atomic heritage.org/history/nevada-test-site-downwinders.

Avtandilashvili, Maia, Stacey L. McComish, George Tabatadze, and Sergei Y. Tolmachev. "USTUR Research: Land of Opportunity." Presentation at EURADOS Annual Meeting. KIT, Karlsruhe, Germany. February 27–March 2, 2017.

Bair, William. "The Radiosensitivity of Alligators." In *Pacific Northwest National Laboratory Annual Report for 1967 to the USAEC Division of Biology and Medicine*, edited by R. C. Thompson, Paulette Teal, and Evelyn Swezea, 1.18–1.20. AEC Research and Development Report. BNWL-714. Richland, WA, 1968.

———. "Interview with Bill Bair." Hanford Oral History Project. Washington State University-Tri-Cities. Richland, WA, August 14, 2013.

Baker, Mike. "A Proud Nuclear Town Grapples with How to Remember the Bomb." *New York Times*. September 10, 2022.

Baker, Peter, and Choe Sang-Hun. "Trump Threatens 'Fire and Fury' against North Korea If It Endangers U.S." *New York Times*. August 8, 2017.

Balayannis, Angeliki. "Toxic Sights: The Spectacle of Hazardous Waste Removal." *Environment and Planning D: Society & Space* 38, no. 4 (2020): 772–90.

Banzhaf, H. Spencer. "Retrospectives: The Cold-War Origins of the Value of Statistical Life." *Journal of Economic Perspectives* 28, no. 4 (2014): 213–26.

Barad, Karen. *Meeting the Universe Halfway: Quantum Physics and the Entanglement of Matter and Meaning*. Durham, NC: Duke University Press, 2007.

Barker, Holly M. *Bravo for the Marshallese: Regaining Control in a Post-Nuclear, Post-Colonial World*. Belmont, CA: Wadsworth, 2013.

———. "Unsettling SpongeBob and the Legacies of Violence on Bikini Bottom." *Contemporary Pacific* 31, no. 2 (2019): 345–79.

Barras, Colin. "This Roman 'Gate to Hell' Killed Its Victims with a Cloud of Deadly Carbon Dioxide." *Science*. February 16, 2018.

Beatty, John. "Masking Disagreement among Experts." *Episteme* 3, nos. 1–2 (2006): 52–67.

Benjamin, Ruha. "Speculative Futures: Envisioning and Creating Social and Reproductive Justice in These Times." Keynote address at University of California at Davis. October 31, 2018.

———. "Race to the Future: Rethinking Innovation, Inequity, and Imagination in Everyday Life." Keynote address at Emerson College, Boston, MA. October 17, 2019.

"Body of Nuclear Scientist Studied after 4-Month Wait." *Oregonian*. September 3, 1979.

Boland, Joseph B. "The Cold War Legacy of Regulatory Risk Analysis: The Atomic Energy Commission and Radiation Safety." University of Oregon. ProQuest Dissertations Publishing, 2002.

Bolman, Brad. "Pig Mentations: Race and Face in Radiobiology." *Isis* 112, no. 4 (2021): 694–716.
Bolton, Matthew Breay. "Human Rights Fallout of Nuclear Detonations: Reevaluating 'Threshold Thinking' in Assisting Victims of Nuclear Testing." *Global Policy* 13, no. 1 (2022): 76–90.
Boyd, M. A. "A Regulatory Perspective on Whether the System of Radiation Protection Is Fit for Purpose." *Annals of the ICRP* 41, nos. 3–4 (2012): 57–63.
Boyd, William. "Genealogies of Risk: Searching for Safety, 1930s–1970s." *Ecology Law Quarterly* 39, no. 4 (2012): 895–987.
Breitenstein, B. D., et al. "The U.S. Transuranium Registry Report on the 241Am Content of a Whole Body." *Health Physics* 49, no. 4 (1985): 559–661.
Breyer, Stephen. *Breaking the Vicious Circle: Toward Effective Risk Regulation.* Cambridge, MA: Harvard University Press, 1993.
Bridgen, Pamela. "Protecting Native Americans through the Risk Assessment Process: A Commentary on 'An Examination of U.S. EPA Risk Assessment Principles and Practices.'" *Integrated Environmental Assessment and Management* 1, no. 1 (2005): 83–85.
Brooks, Antone L. *Low Dose Radiation: The History of the U.S. Department of Energy Research Program.* Pullman: Washington State University Press, 2018.
Brown, David. "Aid to Atomic Weapons Workers." *Washington Post.* July 16, 1999.
Brown, Kate. *Plutopia: Nuclear Families, Atomic Cities, and the Great Soviet and American Plutonium Disasters.* Oxford: Oxford University Press, 2013.
Brown, Wendy. *States of Injury: Power and Freedom in Late Modernity.* Princeton, NJ: Princeton University Press, 1995.
Burgio, Ernesto, Prisco Piscitelli, and Lucia Migliore. "Ionizing Radiation and Human Health: Reviewing Models of Exposure and Mechanisms of Cellular Damage; An Epigenetic Perspective." *International Journal of Environmental Research and Public Health* 15, no. 9 (2018): 1971.
Butler, Judith. "Performative Acts and Gender Constitution: An Essay in Phenomenology and Feminist Theory." *Theatre Journal* 40, no. 4 (1988): 519–31.
———. *Bodies That Matter: On the Discursive Limits of "Sex."* New York: Routledge, 1993.
———. *Frames of War: When Is Life Grievable?* London: Verso, 2009.
Cameron, Kim S., and Marc Levine. *Making the Impossible Possible: Leading Extraordinary Performance—the Rocky Flats Story.* San Francisco: Berrett-Koehler, 2006.
Cartwright, Lisa. "A Cultural Anatomy of the Visible Human Project." In *The Visible Woman: Imaging Technologies, Gender, and Science*, edited by Paula A. Treichler, Lisa Cartwright, and Constance Penley, 21–43. New York: New York University Press, 1998.

Cary, Annette. "Fruit Flies Suspects in Hanford's Contamination." *Tri-City Herald*. October 7, 1998.
———. "DOE Report Long on Horizon." *Tri-City Herald*. February 27, 2000.
———. "Japanese Exchange Student Offers Her Views on Richland High School's Mushroom Cloud Logo." *Seattle Times*. June 10, 2019.
———. "57% of Hanford Nuclear Site Workers Surveyed by Washington State Report Toxic Exposures." *Tri-City Herald*. July 7, 2021.
Centers for Disease Control. "Radioisotope Brief: Plutonium." https://www.cdc.gov/nceh/radiation/emergencies/isotopes/plutonium.htm.
Chen, J., D. M. Sridharan, C. L. Cross, and J. M. Pluth. "Cellular DNA Effects of Radiation and Cancer Risk Assessment in Cells with Mitochondrial Defects." *Journal of Radiation Research and Applied Sciences* 15, no. 1 (2022): 89–94.
Chen, Mel Y. "Toxic Animacies, Inanimate Affections." *GLQ* 17, nos. 2–3 (2011): 265–86.
Clarke, Lee. *Mission Improbable: Using Fantasy Documents to Tame Disaster*. Chicago: University of Chicago Press, 1999.
Cohen, David D. Letter to Linus Pauling. October 29, 1959. Oregon State University Special Collections and Archives Research Center, Ava Helen and Linus Pauling Papers, box 7.012, folder 12.17.
Columbia River Inter-Tribal Fish Commission. *A Fish Consumption Survey of the Umatilla, Nez Perce, Yakama, and Warm Springs Tribes of the Columbia River Basin*. Technical Report 94-3. Portland, OR, 1994.
Columbia Riverkeeper. "Competing Visions for the Future of Hanford." June 2018. https://www.columbiariverkeeper.org/sites/default/files/2018-07/2018.6.1%20Hanford%20Vision%20Report_j_INTERACTIVE.pdf.
———. "Thousands Urge Federal Government to Drop Proposal to Re-classify High-Level Waste at Hanford." November 7, 2018. https://www.columbiariverkeeper.org/news/2018/11/thousands-urge-federal-government-drop-proposal-reclassify-high-level-nuclear-waste.
Confederated Tribes of the Umatilla Indian Reservation. "Treaty of 1855." https://www.ctuir.org/departments/office-of-legal-counsel/codes-statutes-laws/treaty-of-1855/.
"The Contaminators." *Playboy*. October 1959. Oregon State University Special Collections and Archives Research Center, Ava Helen and Linus Pauling Papers, box 7.012, folder 12.6.
Council on Environmental Quality. "Update to the Regulations Implementing the Procedural Provisions of the National Environmental Policy Act." *Federal Register*. 85 FR 43304. July 16, 2020.
Cousins, Norman. "The Debate Is Over." *Saturday Review*. April 4, 1959. Oregon State University Special Collections and Archives Research Center, Ava Helen and Linus Pauling Papers, box 7.003, folder 3.24.

Coviello, Peter. "Apocalypse from Now On." In *Queer Frontiers: Millennial Geographies, Genders, and Generations*, edited by Joseph Allen Boone, 39–63. Madison: University of Wisconsin Press, 2000.
Cram, Shannon. "Escaping S-102: Waste, Illness, and the Politics of Not Knowing." *International Journal of Science in Society* 2, no. 1 (2011): 243–52.
———. "Becoming Jane: The Making and Unmaking of Hanford's Nuclear Body." *Environment and Planning D: Society and Space* 33, no. 5 (2015): 796–812.
———. "Wild and Scenic Wasteland: Conservation Politics in the Nuclear Wilderness." *Environmental Humanities* 7, no. 1 (2016): 89–105.
———. "Living in Dose: Nuclear Work and the Politics of Permissible Exposure." *Public Culture* 28, no. 3 (2016): 519–39.
Crane, Johanna Tayloe. *Scrambling for Africa: AIDS, Expertise, and the Rise of American Global Health Science*. Ithaca, NY: Cornell University Press, 2013.
Creager, Angela. "Radiation, Cancer, and Mutation in the Atomic Age." *Historical Studies in the Natural Sciences* 45, no. 1 (2015): 14–48.
Cristy, M. "Mathematical Phantoms Representing Children of Various Ages for Use in Estimates of Internal Dose." ORNL/NUREG/TM-367. Oak Ridge, TN: Oak Ridge National Laboratory, 1980.
———. "Representative Breast Size of Reference Female." *Health Physics* 43 (1982): 930–32.
———. "Calculation of Annual Limits of Intake of Radionuclides by Workers: Significance of Breast as an Explicitly Represented Tissue." *Health Physics* 46, no. 2 (1984): 283–91.
———. "Reference Man Anatomical Model." Conference presentation, Health Physics Society Summer School on Internal Dosimetry. Davis, CA. June 6, 1994. https://www.osti.gov/biblio/10186060.
Cristy, M., and K. F. Eckerman. "Specific Absorbed Fractions of Energy at Various Ages from Internal Photon Sources." ORNL/TM-8381. Oak Ridge, TN: Oak Ridge National Laboratory, 1987.
Crow, James. "Quarrelling Geneticists and a Diplomat." *Genetics* 140, no. 2 (1995): 421–26.
"Curtiss-Wright EST Group." *Radwaste Solutions* 25, no. 1 (2018): 13.
de la Torre III, Pedro. "Unmaking Wastelands: Inheriting Waste, War, and Futures at the Hanford Site." Rensselaer Polytechnic Institute. Pro Quest Dissertations Publishing, 2017.
DeLoughrey, Elizabeth. 2013. "The Myth of Isolates: Ecosystem Ecologies in the Nuclear Pacific." *Cultural Geographies* 20 (2): 167–84.
Dillon, Lindsey. "Race, Waste, and Space: Brownfield Redevelopment and Environmental Justice at the Hunters Point Shipyard." *Antipode* 46, no. 5 (2014): 1205–21.

Dirkes, R. L., R. W. Hanf, and T. M. Poston. *Hanford Site Environmental Report for Calendar Year 1998.* Prepared for the U.S. Department of Energy. PNNL-12088. September 1999.

"Discussions with Deb: The UPPU Club." *Cold War Patriots* (blog). July 16, 2021. https://coldwarpatriots.org/blog/discussions-with-deb-the-uppu-club/.

Dubrova, Yuri E., and Elena I. Sarapultseva. "Radiation-Induced Transgenerational Effects in Animals." *International Journal of Radiation Biology* 98, no. 6 (2022): 1047–53.

Eckerman, K. F. and Jeffrey Clair Ryman. *External Exposure to Radionuclides in Air, Water, and Soil: Exposure-to-Dose Coefficients for General Application, Based on the 1987 Federal Radiation Protection Guidance.* Washington, D.C.: Office of Radiation and Indoor Air, U.S. Environmental Protection Agency, 1993.

Eckerman, K. F., and M. Cristy. *The Reference Individual of Radiation Protection.* No. CONF-9507235--1. Oak Ridge, TN: Oak Ridge National Laboratory, 1995.

Edwards, Paul N. *The Closed World: Computers and the Politics of Discourse in Cold War America.* Cambridge, MA: MIT Press, 1997.

Eisenbud, Merril. "The Risks." In *America Faces the Nuclear Age*, edited by Johnson E. Fairchild and David Landman, 91–106. New York: Sheridan House, 1961.

Employee Radiation Hazards and Workmen's Compensation: Hearings before the Subcommittee on Research and Development of the Joint Committee on Atomic Energy. 86th Cong., 1st Sess. March 10–19, 1959.

Encyclopedia Britannica. S.v. "Valkyrie." November 28, 2022. https://www.britannica.com/topic/Valkyrie-Norse-mythology.

"Endeavor Robotics." *Radwaste Solutions* 25, no. 1 (2018): 17.

Energy Employees Occupational Illness Compensation Program; Are We Fulfilling the Promise We Made to These Cold War Veterans When We Created This Program?: Hearing before the Subcommittee on Immigration, Border Security, and Claims. 109th Cong., 2nd Sess. May 4, 2006.

Erickson, Paul, Judy L. Klein, Lorraine Daston, Rebecca Lemov, Thomas Sturm, and Michael D. Gordin, eds. *How Reason Almost Lost Its Mind: The Strange Career of Cold War Rationality.* Chicago: University of Chicago Press, 2013.

Fanon, Frantz. *Black Skin, White Masks.* Translated by Charles Lam Markmann. New York: Grove Press, 1967.

Farah, Jad, David Broggio, and Didier Franck. "Creation and Use of Adjustable 3D Phantoms: Application for the Lung Monitoring of Female Workers." *Health Physics* 99, no. 5 (2010): 649–61.

Filler, Mary. Letter to Linus Pauling. February 3, 1960. Oregon State University Special Collections and Archives Research Center, Ava Helen and Linus Pauling Papers, box 7.013, folder 13.9.

Fletcher, Meg. "Proposal for New Benefits." *Business Insurance* 34, no. 18 (2000): 2.
Flynn, Michael. "A Debt Long Overdue." *Bulletin of the Atomic Scientists* 57, no. 4 (2001): 38–48.
Fogleman, Valerie M. "Worst Case Analyses: A Continued Requirement under the National Environmental Policy Act?" *Columbia Journal of Environmental Law* 13, no. 1 (1987): 53.
Folkers, Cynthia. "Disproportionate Impacts of Radiation Exposure on Women, Children, and Pregnancy: Taking Back Our Narrative." *Journal of the History of Biology* 54, no. 1 (2021): 31–66.
Foresman, Mrs. Robert. Letter to Linus Pauling. September 22, 1959. Oregon State University Special Collections and Archives Research Center, Ava Helen and Linus Pauling Papers, box 7.012, folder 12.17.
Foucault, Michel. *Discipline and Punish: The Birth of the Prison*. New York: Vintage Books, 1979.
———. *The Birth of the Clinic: An Archaeology of Medical Perception*. New York: Vintage Books, 1994.
———. *Security, Territory Population: Lectures at the College de France, 1977–1978*. New York: Picador, 2004.
French, Norman R. 1960. "Strontium-90 in Ecuador." *Science* 131, no. 3417 (1960): 1889–90. Oregon State University Special Collections and Archives Research Center, Ava Helen and Linus Pauling Papers, box 7.013, folder 13.10.
Gabbert, Elisa. *The Unreality of Memory: and Other Essays*. New York: Farrar, Straus and Giroux, 2020.
Garcia, Angela. *The Pastoral Clinic: Addiction and Dispossession along the Rio Grande*. Berkeley: University of California Press, 2010.
"Genetics Panel, Second Meeting. Chicago, IL, February 5 and 6, 1956." CalTech, George Wells Beadle Papers, box 17, folder 17.1.
George, Marlene. "Review of the Remedial Investigation/Feasibility Study and Proposed Plan for the 100-BC-1, BC-2 and BC-5 Operable Units, DOE/RL-2010-96 and DOE/RL-2016-43, Draft A." January 11, 2018. https://www.columbiariverkeeper.org/sites/default/files/2019-11/YN%20final%20comments%20-1-11-2018%20re%20100-B-C%20RIFS-PP-DOE-RL-2010-96%20-DOE-RL-2016-43-Draft%20A%20%283%29%20%281%29.pdf.
Gephart, R. E. *Hanford: A Conversation about Nuclear Waste and Cleanup*. Columbus, OH: Battelle Press, 2003.
Gephart, Roy. "A Short History of Waste Management at the Hanford Site." *Physics and Chemistry of the Earth* 35, no. 6 (2010): 298–306.
Gerber, Michele Stenehjem. *Legend and Legacy: Fifty Years of Defense Production at the Hanford Site*. Richland, WA: Westinghouse Hanford, 1992.
———. "All-Out Effort Stems Spreads of Contamination." *Hanford Reach*. October 19, 1998.

———. *On the Home Front: The Cold War Legacy of the Hanford Nuclear Site*. Lincoln: University of Nebraska Press, 2002.

Gochfeld, Michael, and Joanna Burger. "Disproportionate Exposures in Environmental Justice and Other Populations: The Importance of Outliers." *American Journal of Public Health* 101, no. S1 (2011): S53–S63.

Gómez, Hernán F., Dominic A. Borgialli, Mahesh Sharman, Keneil K. Shah, Anthony J. Scolpino, James M. Oleske, and John D. Bogden. "Blood Lead Levels of Children in Flint, Michigan: 2006–2016." *Journal of Pediatrics* 197 (2018): 158–64.

Goodman, Michael. "National Radiation Health Standards—A Study in Scientific Decision Making." *Atomic Energy Law Journal* 6, no. 3 (1964): 217–73.

Goodwin, Irwin. "Clinton Apologizes for Cold War's Radiation Experiments, Which Panel Says 'Created a Legacy of Distrust' in Science." *Physics Today* 48, no. 11 (1995): 70.

Gordon, Avery. *Ghostly Matters: Haunting and the Sociological Imagination*. Minneapolis: University of Minnesota Press, 2008.

Grandia, Liza. "Carpet Bombings: A Drama of Chemical Injury in Three Acts." *Catalyst: Feminism, Theory, Technoscience* 6, no. 1 (2020): 1–8.

———. "Toxic Gaslighting: On the Ins and Outs of Pollution." *Engaging Science, Technology, and Society* 6 (2020): 486–513.

Guthman, Julie. *Weighing In: Obesity, Food Justice, and the Limits of Capitalism*. Berkeley: University of California Press, 2011.

Hamblin, Jacob Darwin. "'A Dispassionate and Objective Effort': Negotiating the First Study on the Biological Effects of Atomic Radiation." *Journal of the History of Biology* 40, no. 1 (2007): 147–77.

———. *Poison in the Well: Radioactive Waste in the Oceans at the Dawn of the Nuclear Age*. New Brunswick, NJ: Rutgers University Press, 2008.

———. *Arming Mother Nature: The Birth of Catastrophic Environmentalism*. Oxford: Oxford University Press, 2013.

Hamilton, Kevin, and Ned O'Gorman. "Seeing Experimental Imperialism in the Nuclear Pacific." *Media+Environment* 3, no. 1 (2021): 1–23.

Hanford Advisory Board. "HAB Consensus Advice #294. Subject: Hanford Site Budget." November 13, 2017.

Hanford Challenge, Natural Resources Defense Council, and Columbia Riverkeeper. "Comments on Draft Waste Incidental to Reprocessing Evaluation for Closure of Waste Management Area C at the Hanford Site." November 7, 2018. https://static1.squarespace.com/static/568adf4125981deb769d96b2/t/608352e267faf3797471c50a/1619219171924/NRDC+et+al+Draft+WIR+Comment+%26+Attachments+sm.pdf.

Hanford Natural Resource Damage Assessment Injury Assessment Plan. Richland, WA: Hanford Natural Resource Trustees, 2013.

Hansson, Sven Ove. "Should We Protect the Most Sensitive People?" *Journal of Radiological Protection* 29, no. 2 (2009): 211–18.

Haraway, Donna. *Simians, Cyborgs, and Women: The Reinvention of Nature.* New York: Routledge, 1991.

Harney, Stefano, and Fred Moten. *The Undercommons: Fugitive Planning and Black Study.* Wivenhoe, UK: Minor Compositions, 2013.

Harper, Barbara, Brian Flett, Stuart Harris, Corn Abeyta, and Fred Kirschner. "Response to Letter to the Editor." *Risk Analysis* 23, no. 5 (2003): 861–64.

Harper, Barbara, Anna Harding, Stuart Harris, and Patricia Berger. "Subsistence Exposure Scenarios for Tribal Applications." *Human and Ecological Risk Assessment* 18, no. 4 (2012): 810–31.

Harper, Barbara L., Anna K. Harding, Therese Waterhous, and Stuart G. Harris. *Traditional Tribal Subsistence Exposure Scenario and Risk Assessment Guidance Manual.* Corvallis: Oregon State University, 2007.

Harris, S. G., and B. L. Harper. "Exposure Scenario for CTUIR Traditional Subsistence Lifeways." Department of Science and Engineering. Confederated Tribes of the Umatilla Indian Reservation. Pendleton, OR, 2004.

Harrison, Jill. *From the Inside Out: The Fight for Environmental Justice within Governmental Agencies.* Cambridge, MA: MIT Press, 2019.

Hartman, Saidiya. "Venus in Two Acts." *Small Axe: A Journal of Criticism* 12, no. 2 (2008): 1–14.

Havlick, David. "Logics of Change for Military-to-Wildlife Conversions in the United States." *GeoJournal* 69, no. 3 (2007): 151–64.

Havlick, David G. "Disarming Nature: Converting Military Lands to Wildlife Refuges." *Geographical Review* 101, no. 2 (2011): 183–200.

———. "Opportunistic Conservation at Former Military Sites in the United States." *Progress in Physical Geography* 38, no. 3 (2014): 271–85.

Hecht, Gabrielle. *Being Nuclear: Africans and the Global Uranium Trade.* Cambridge, MA: MIT Press, 2012.

———. "The Work of Invisibility: Radiation Hazards and Occupational Health in South African Uranium Production." *International Labor and Working-Class History* 81, no. 81 (2012): 94–113.

Hegenbart, L., Y. H. Na, J. Y. Zhang, M. Urban, and X. George Xu. "A Monte Carlo Study of Lung Counting Efficiency for Female Workers of Different Breast Sizes Using Deformable Phantoms." *Physics in Medicine & Biology* 53, no. 19 (2008): 5527–38.

Heinzerling, Lisa. "The Rights of Statistical People." *Harvard Environmental Law Review* 24, no. 1 (2000): 189–207.

———. "Knowing Killing and Environmental Law." *New York University Environmental Law Journal* 14, no. 3 (2006): 521–736.

———. Session IV: "Does the Law Tend to Favor Identified over Statistical People?" Harvard University, 7th Annual Program in Ethics and Health Conference. May 1, 2012.

———. "Statistical Lives in Environmental Law." In *Identified versus Statistical Lives: An Interdisciplinary Perspective*, edited by Glenn Cohen,

Normal Daniels, and Nir Eyal, 174–81. Oxford: Oxford University Press, 2015.
Heller, Stephen R., John M. McGuire, and William L. Budde. "Trace Organics by GC/MS [Gas Chromatography/Mass Spectrometry]." *Environmental Science & Technology* 9, no. 3 (1975): 210–13.
Helton, Laura, Justin Leroy, Max A. Mishler, Samantha Seeley, and Shauna Sweeney. "The Question of Recovery." *Social Text* 33, no. 4 (2015): 1–18.
Hepler-Smith, Evan. "Molecular Bureaucracy: Toxicological Information and Environmental Protection." *Environmental History* 24, no. 3 (2019): 534–60.
Hickman, D. P., and N. Cohen. "Reconstruction of a Human Skull Calibration Phantom Using Bone Sections from an ^{241}Am Exposure Case." *Health Physics* 55, no. 1 (1988): 59–65.
Higuchi, Toshihiro. *Political Fallout: Nuclear Weapons Testing and the Making of a Global Environmental Crisis.* Stanford, CA: Stanford University Press, 2020.
Hill, Corinne E., J. P. Myers, and Laura N. Vandenberg. "Nonmonotonic Dose–Response Curves Occur in Dose Ranges That Are Relevant to Regulatory Decision-Making." *Dose-Response* 16, no. 3 (2018): 1–4.
Hoover, Elizabeth. *The River Is in Us: Fighting Toxics in a Mohawk Community.* Minneapolis: University of Minnesota Press, 2017.
H.R. 1208, to Establish the Manhattan Project National Historical Park in Oak Ridge, TN, Los Alamos, NM, and Hanford, WA: Legislative Hearing before the Subcommittee on Public Lands and Environmental Regulation of the Committee on Natural Resources, U.S. House of Representatives. 113th Cong., 1st Sess. April 12, 2013.
Hurley, Jessica. *Infrastructures of Apocalypse: American Literature and the Nuclear Complex.* Minneapolis: University of Minnesota Press, 2020.
Hutt, Peter B. "Use of Quantitative Risk Assessment in Regulatory Decision-making under Federal Health and Safety Statutes." In *Risk Quantitation and Regulatory Policy*, edited by David G. Hoel, Richard A. Merrill, and Frederica P. Perera, 15–29. Cold Spring Harbor, NY: Cold Spring Harbor Laboratory, 1985.
International Atomic Energy Agency. "Intercalibration of In Vivo Counting Systems Using an Asian Phantom." IAEA-TECDOC-1334. Radiation Monitoring and Protection Section. Vienna, Austria, 2003.
International Commission on Radiological Protection. *Report of Task Group on Reference Man.* ICRP Publication 23. Oxford: Pergamon Press, 1975.
——. *Basic Anatomical and Physiological Data for Use in Radiological Protection: Reference Values.* New York: Pergamon Press, 2003.
——. "The 2007 Recommendations of the International Commission on Radiological Protection. ICRP Publication 103." *Annals of the ICRP* 37, nos. 2–4 (2007): 1–332.

———. "Environmental Protection—the Concept and Use of Reference Plants and Animals. ICRP Publication 108." *Annals of the ICRP* 38, nos. 4–6 (2008): 1–242.

"Is This What It's Coming To?" *New York Times*. July 5, 1962. Oregon State University Special Collections and Archives Research Center, Ava Helen and Linus Pauling Papers, box 7.014, folder 14.11.

Iversen, Kristen. *Full Body Burden: Growing up in the Nuclear Shadow of Rocky Flats*. New York: Crown, 2012.

Jacobs, Robert A. *Nuclear Bodies the Global Hibakusha*. New Haven, CT: Yale University Press, 2022.

Jain, S. Lochlann. *Injury: The Politics of Product Design and Safety Law in the United States*. Princeton, NJ: Princeton University Press, 2006.

———. *Malignant: How Cancer Becomes Us*. Berkeley: University of California Press, 2013.

James, Tony. "DOE's U.S. Transuranium and Uranium Registries (USTUR): Studying Occupational Exposures to Plutonium from Beginning to End." Special presentation to DOE/EH, December 14, 2005.

Jim, Russell. Interview by Cynthia Kelly, Tom Zannes, and Thomas E. Marceau. Atomic Heritage Foundation. "Voices of the Manhattan Project." Hanford, WA. https://ahf.nuclearmuseum.org/voices/oral-histories/russell-jims-interview/.

Johnson, A. R., J. G. Cardiff, R. F. Giddings, et al. "An Integrated Biological Control System at Hanford." Presentation to the Waste Management Symposia, Phoenix, AZ, March 8, 2010.

Johnson, A. R., and M. J. Elsen. "An Integrated Biological Control System at Hanford." Presentation to the Hanford Advisory Board. Richland, WA, April 15, 2010.

Joint Legislative Audit Committee. "The Cancer Incidence among Workers at the Lawrence Livermore Laboratory: A Synthesis of Expert Reviews of the Study." Sacramento, CA, 1980.

Jolly, J. Christopher. "Thresholds of Uncertainty: Radiation and Responsibility in the Fallout Controversy." Oregon State University. ProQuest Dissertations Publishing, 2004.

Jordanova, L. J. *Sexual Visions: Images of Gender in Science and Medicine Between the Eighteenth and Twentieth Centuries*. Madison: University of Wisconsin Press, 1989.

Jose, Donald, and David Wiedis. "Preparing the Health Physicist to Testify at Deposition." *Radiological Safety Officer Magazine* 8, no. 1 (2003): 23–29.

Joyce, Rosemary. *The Future of Nuclear Waste: What Art and Archaeology Can Tell Us about Securing the World's Most Hazardous Material*. Oxford: Oxford University Press, 2020.

Judicial Council of California. "Civil Jury Instructions." 2020 edition. California Civil Code 401. "Basic Standard of Care." https://www.justia.com/trials-litigation/docs/caci/400/401/.

Kandic, Slavica, Susanne J. Tepe, Ewan W. Blanch, Shamali De Silva, Hannah G. Mikkonen, and Suzie M. Reichman. "Quantifying Factors Related to Urban Metal Contamination in Vegetable Garden Soils of the West and North of Melbourne, Australia." *Environmental Pollution* 251 (2019): 193–202.

Kephart, Gary Steven. "An Arm Phantom for in Vivo Determination of Americium-241 in Bone." Master's thesis, University of Washington, 1987.

Kiarashi, Nooshin, Adam C. Nolte, Gregory M. Sturgeon, William P. Segars, Sujata V. Ghate, Loren W. Nolte, Ehsan Samei, and Joseph Y. Lo. "Development of Realistic Physical Breast Phantoms Matched to Virtual Breast Phantoms Based on Human Subject Data." *Medical Physics* 42, no. 7 (2015): 4116–26.

Kimura, Aya Hirata. *Radiation Brain Moms and Citizen Scientists: The Gender Politics of Food Contamination after Fukushima.* Durham, NC: Duke University Press, 2016.

Knibbe, Morgan. "The Atomic Soldiers." *New York Times Op-Docs.* Season 6. March 16, 2019. https://www.nytimes.com/2019/02/12/opinion/atomic-soldiers.html.

Kosek, Jake. "Aggregate Modernities: A Critical Natural History of Contemporary Algorithms." In *Other Geographies: The Influence of Michael Watts*, edited by Sharad Chari, Susanne Friedberg, Vinay Gidwani, Jesse Ribot, and Wendy Wolford, 63–78. Newark, NJ: Wiley-Blackwell, 2017.

Kramer, R., and G. Drexler. "Representative Breast Size of Reference Female." *Health Physics* 40 (1981): 914.

Kramer, R., G. Williams, and G. Drexler. "Reply to M. Cristy." *Health Physics* 43 (1982): 932–35.

Kramer, G. H., B. M. Hauck, and S. A. Allen. "Comparison of the LLNL and JAERI Torso Phantoms Using Ge Detectors and Phoswich Detectors." *Health Physics* 74, no. 5 (1998): 594–601.

Kramer, Gary, et al. "Comparison of Two Leg Phantoms Containing ^{241}Am in Bone." *Health Physics* 101, no. 3 (2011): 248–58.

Krupar, Shiloh R. "Alien Still Life: Distilling the Toxic Logics of the Rocky Flats National Wildlife Refuge." *Environment and Planning D: Society & Space* 29, no. 2 (2011): 268–90.

———. "The Biopsic Adventures of Mammary Glam: Breast Cancer Detection and the Practice of Cancer Glamor." *Social Semiotics* 22, no. 1 (2012): 47–82.

———. *Hot Spotter's Report: Military Fables of Toxic Waste.* Minneapolis: University of Minnesota Press, 2013.

———. "The Biomedicalisation of War and Military Remains: US Nuclear Worker Compensation in the 'Post–Cold War.'" *Medicine, Conflict, and Survival* 29, no. 2 (2013): 111–39.

Kuletz, Valerie. *The Tainted Desert: Environmental Ruin in the American West.* New York: Routledge, 1998.

Kulp, Lawrence, and Arthur Schulert. "Strontium-90 in Man V." *Science* 135, no. 3516 (1962): 619–32. Oregon State University Special Collections and

Archives Research Center, Ava Helen and Linus Pauling Papers, box 7.014, folder 14.10.

Kulp, Lawrence, Arthur Schulert, and Elizabeth J. Hodges. "Strontium-90 in Man III." *Science* 129, no. 3358 (1959): 1249–55. Oregon State University Special Collections and Archives Research Center, Ava Helen and Linus Pauling Papers, box 7.001, folder 11.12.

Kysar, Douglas A. *Regulating from Nowhere: Environmental Law and the Search for Objectivity*. New Haven, CT: Yale University Press, 2010.

Kunimoto, Namiko. *The Stakes of Exposure: Anxious Bodies in Postwar Japanese Art*. Minneapolis: University of Minnesota Press, 2017.

LaDuke, Winona. "Uranium Mining, Native Resistance, and the Greener Path." *Orion* 28, no. 1 (2009): 22.

Lee, Gary. "Letting the Nation in on Decades of Secrets." *Washington Post*. March 31, 1994.

Lepawsky, Josh. *Reassembling Rubbish: Worlding Electronic Waste*. Cambridge, MA: MIT Press, 2018.

Liboiron, Max. *Pollution Is Colonialism*. Durham, NC: Duke University Press, 2021.

Liboiron, Max, Manuel Tironi, and Nerea Calvillo. "Toxic Politics: Acting in a Permanently Polluted World." *Social Studies of Science* 48, no. 3 (2018): 331–49.

Lindee, M. Susan. *Suffering Made Real: American Science and the Survivors at Hiroshima*. Chicago: University of Chicago Press, 1994.

Linton, Etta. Letter to Linus Pauling. August 13, 1962. Oregon State University Special Collections and Archives Research Center, Ava Helen and Linus Pauling Papers, box 7.014, folder 14.7.

Lombardo, Pasquale Alessandro, Anne Laure Lebacq, and Filip Vanhavere. "Creation of Female Computational Phantoms for Calibration of Lung Counters." *Radiation Protection Dosimetry* 170, nos. 1–4 (2016): 369–72.

Lynch, T. P. "In Vivo Monitoring Program Manual." HNF-55649. Mission Support Alliance (U.S. Department of Energy). Richland, WA, 2014.

Lynch, Timothy P. "In Vivo Radiobioassay and Research Facility." *Health Physics* 100, no. 2 (2011): 35–40.

———. "In Vivo Monitoring Program Manual." PNL-MA-574. Pacific Northwest National Laboratory (U.S. Department of Energy). Richland, WA, 2011.

Makhijani, Arjun, Brice Smith, and Michael Thorne. "Science for the Vulnerable: Setting Radiation and Multiple Exposure Environmental Health Standards to Protect Those Most at Risk." Takoma Park, MD: Institute for Energy and Environmental Research, 2006.

Malone, Melanie. "Seeking Justice, Eating Toxics: Overlooked Contaminants in Urban Community Gardens." *Agriculture and Human Values* 39, no. 1 (2022): 165–84.

Manohari, M. "Simulation of In-Vivo Monitors and VOXEL Phantoms for Establishing Calibration Factors." PhD dissertation, Homi Bhabha National Institute, Mumbai, India, 2014.

Masco, Joseph. "Mutant Ecologies: Radioactive Life in Post–Cold War New Mexico." *Cultural Anthropology* 19, no. 4 (2004): 517–50.
———. *The Nuclear Borderlands: The Manhattan Project in Post–Cold War New Mexico.* Princeton, NJ: Princeton University Press, 2006.
———. "'Survival Is Your Business': Engineering Ruins and Affect in Nuclear America." *Cultural Anthropology* 23, no. 2 (2008): 361–98.
———. "'Sensitive but Unclassified': Secrecy and the Counterterrorist State." *Public Culture* 22, no. 3 (2010): 433–63.
———. *The Theater of Operations: National Security Affect from the Cold War to the War on Terror.* Durham, NC: Duke University Press, 2014.
———. "The Age of Fallout." *History of the Present* 5, no. 2 (2015): 137–68.
———. *The Future of Fallout, and Other Episodes in Radioactive World-Making.* Durham, NC: Duke University Press, 2021.
Maupin, Ann M. Letter to Linus Pauling. December 2, 1959. Oregon State University Special Collections and Archives Research Center, Ava Helen and Linus Pauling Papers, box 7.012, folder 12.17.
McBride, Stewart. "New Life for Nuclear City." *New York Times.* January 18, 1981.
McInroy, J. F., H. A. Boyd, B. C. Eutsler, and D. Romero. "The U.S. Transuranium Registry Report of the 241Am Content of a Whole Body: Part IV, Preparation and Analysis of the Tissues and Bones." *Health Physics* 49, no. 4 (1985): 587.
McInroy, J. F., R. L. Kathren, R. E. Toohey, M. J. Swint, and B. D. Breitenstein. "Postmortem Tissue Contents of 241Am in a Person with a Massive Acute Exposure." *Health Physics* 69, no. 3 (1995): 318–23.
McClellan, R. O., and C. R. Watson. "Radiation Dosimetry of Cs 137 in Sheep Evaluated with Thermoluminscent Dosimeters." In *Hanford Biology Research Annual Report for 1964.* Pacific Northwest Laboratory (U.S. Department of Energy), 1965.
Michaels, David. *Doubt Is Their Product: How Industry's Assault on Science Threatens Your Health.* Oxford: Oxford University Press, 2008.
Mika, Marissa. *Africanizing Oncology: Creativity, Crisis, and Cancer in Uganda.* Columbus: Ohio University Press, 2021.
Miller, Fred. *U.S. Transuranium and Uranium Registries Analytical Procedure Manual.* Method Number: USTUR 1000. Richland, WA, 2011. https://s3.wp.wsu.edu/uploads/sites/1058/2016/10/USTUR_1000.pdf.
Miłosz, Czesław. *The Collected Poems, 1931–1987.* New York: Ecco Press, 1988.
Moore, Lisa Jean, and Adele E. Clarke. "The Traffic in Cyberanatomies: Sex/Gender/Sexualities in Local and Global Formations." *Body & Society* 7, no. 1 (2001): 57–96.
Morgan, Jennifer L. *Laboring Women: Reproduction and Gender in New World Slavery.* Philadelphia: University of Pennsylvania Press, 2004.
Mosby's Medical Dictionary. 7th ed. St. Louis, MO: Mosby/Elsevier, 2006.
Muller, Herman. "The Radiation Danger." *Colorado Quarterly* 6 (1958): 229–54.

Murphy, Kim. "Government Finally Hears a Nuclear Town's Horrors." *Los Angeles Times*. February 5, 2000.

Murphy, M. *Sick Building Syndrome and the Problem of Uncertainty: Environmental Politics, Technoscience, and Women Workers*. Durham, NC: Duke University Press, 2006.

———. "Unsettling Care: Troubling Transnational Itineraries of Care in Feminist Health Practices." *Social Studies of Science* 45, no. 5 (2015): 717–37.

———. "Alterlife and Decolonial Chemical Relations." *Cultural Anthropology* 32, no. 4 (2017): 494–503.

———. *The Economization of Life*. Durham, NC: Duke University Press, 2017.

———. "What Can't a Body Do?" *Catalyst: Feminism, Theory, Technoscience* 3, no. 1 (2017): 1–15.

Nalder, Eric. "The Plot to Get Ed Bricker—Hanford Whistle-Blower Was Tracked, Harassed, Files Show." *Seattle Times*. July 30, 1990.

Narendran, Nadia, Lidia Luzhna, and Olga Kovalchuk. "Sex Difference of Radiation Response in Occupational and Accidental Exposure." *Frontiers in Genetics* 10 (2019): 1–11.

Nash, Linda. "From Safety to Risk: The Cold War Contexts of American Environmental Policy." *Journal of Policy History* 29, no. 1 (2017): 1–33.

National Academy of Sciences. "The Biological Effects of Atomic Radiation: Summary Reports." Washington, D.C.: National Research Council, 1956.

———. "Minutes of Meeting: Study Group on Dispersal and Disposal of Radioactive Wastes." Washington, D.C., February 23, 1956, 7–8.

"National Academies Panel Says More Work Is Needed on Hanford Waste Report." *Exchange Monitor*. November 9, 2022.

National Aeronautics and Space Administration (NASA). "R5." September 23, 2015. https://www.nasa.gov/feature/r5/.

National Institute for Occupational Safety and Health. "Review of Hanford Tank Farm Worker Safety and Health Programs." Centers for Disease Control and U.S. Department of Health and Human Services. November 28, 2016.

National Research Council. *Toxicity Testing: Strategies to Determine Needs and Priorities*. Washington, D.C.: National Academies Press, 1984.

———. *Long-Term Institutional Management of U.S. Department of Energy Legacy Waste Sites*. Washington, D.C.: National Academies Press, 2000.

———. *Health Risks from Exposure to Low Levels of Ionizing Radiation: BEIR VII Phase 2*. Washington, D.C.: National Academies Press, 2006.

The Nature of Radioactive Fallout and Its Effects on Man: Hearings before the Special Subcommittee on Radiation of the Joint Committee on Atomic Energy on the Nature of Radioactive Fallout and Its Effects on Man. 85th Cong., 1st Sess. 1957.

Nelson, I. C., K. R. Heid, P. A. Fuqua, and T. D. Mahony. "Plutonium in Autopsy Tissue Samples." *Health Physics* 22, no. 6 (1972): 925–30.

Newell, R. R. "Genetic Injuries." *Radiology* 69, no. 1 (1957): 111–14.

Newton, C. E., K. R. Heid, H. V. Larson, I. C. Nelson, and Pacific Northwest Laboratory, publisher. "Tissue Sampling for Plutonium through an Autopsy Program." Battelle Memorial Institute, Pacific Northwest Laboratory. Richland, WA, 1966.

Nez Perce Tribe. "History." https://nezperce.org/about/history/.

Niles, Ken. "The Hanford Cleanup: What's Taking So Long?" *Bulletin of the Atomic Scientists* 70 (2014): 37–48.

———. "Comments on Draft Waste Incidental to Reprocessing Evaluation for Closure of Waste Management Area C at the Hanford Site." October 4, 2018. https://static1.squarespace.com/static/568adf4125981deb769d96b2/t/5bc8f7a624a6949d51e08835/1539897254716/2018-10-4-ODOE-Comments-WIR-Proposal.pdf.

Nixon, Rob. *Slow Violence and the Environmentalism of the Poor*. Cambridge, MA: Harvard University Press, 2011.

Nogueira, P., and W. Ruhm. "Person-Specific Calibration of a Partial Body Counter Used for Individualized ^{241}Am Skull Measurements." *Journal of Radiological Protection* 40 (2020): 1362–89.

Nuclear Testing Program in the Marshall Islands: Hearing before the Committee on Energy and Natural Resources, U.S. Senate. on Effects of U.S. Nuclear Testing Program in the Marshall Islands. 109th Cong., 1st Sess. July 19, 2005.

Oakes, Guy. *The Imaginary War: Civil Defense and American Cold War Culture*. Oxford: Oxford University Press, 1994.

Odum, Eugene P. *Fundamentals of Ecology*. Philadelphia: Saunders, 1959.

Olson, Mary. "Disproportionate Impact of Radiation and Radiation Regulation." *Interdisciplinary Science Reviews* 44, no. 2 (2019): 131–39.

Oregon Department of Energy. "Hanford Groundwater." https://www.oregon.gov/energy/safety-resiliency/Pages/Hanford-Groundwater.aspx.

Orr, Jackie. *Panic Diaries: A Genealogy of Panic Disorder*. Durham, NC: Duke University Press, 2006.

Oxford English Dictionary. S.v. "Phantom." Oxford: Oxford University Press, 2000.

Pacific Northwest National Laboratory. "How Is Groundwater Cleanup Progressing?" https://phoenix.pnnl.gov/phoenix/apps/remediation/index.html.

Palmer, Harvey, et al. "Radioactivity Measurements in Alaskan Eskimos in 1963." *Science* 144, no. 3620 (1964): 859–60. Oregon State University Special Collections and Archives Research Center, Ava Helen and Linus Pauling Papers, box 7.006, folder 6.3.

Paquet, F., G. Etherington, M. R. Bailey, R. W. Leggett, J. Lipsztein, W. Bolch, K. F. Eckerman, and J. D. Harrison. "ICRP Publication 130: Occupational Intakes of Radionuclides, Part 1." *Annals of the ICRP* 44, no. 2 (2015): 5–188.

Parascandola, Mark J. "Compensating for Cold War Cancers." *Environmental Health Perspectives* 110, no. 7 (2002): A404–7.

Parr, Joy. *Sensing Changes: Technologies, Environments, and the Everyday, 1953–2003*. Vancouver: University of British Columbia Press, 2010.
Parshley, Lois. "Cold War, Hot Mess." *Virginia Quarterly Review* 97, no. 3 (2021): 46–71.
Perkins, Tom. "'Forever Chemicals' Detected in All Umbilical Cord Blood in 40 Studies." *Guardian*. September 23, 2022.
Petersen, Gary. "Interview with Gary Petersen." Hanford Oral History Project. Washington State University-Tri-Cities. Richland, WA, June 5, 2014.
Peterson, Val. "Panic, the Ultimate Weapon?" *Colliers*. August 21, 1953.
Petryna, Adriana. *Life Exposed: Biological Citizens after Chernobyl*. Princeton, NJ: Princeton University Press, 2013.
Pfanz, Hardy, Galip Yüce, Ahmet H Gulbay, and Ali Gokgoz. "Deadly CO_2 Gases in the Plutonium of Hierapolis (Denizli, Turkey)." *Archaeological and Anthropological Sciences* 11, no. 4 (2018): 1359–71.
Piepzna-Samarasinha, Leah Lakshmi. *The Future Is Disabled: Prophecies, Love Notes, and Mourning Songs*. Vancouver, BC: Arsenal Pulp Press, 2022.
Price, John. "Hanford Cleanup Priorities Public Meeting." Presentation at the Hanford Site Cleanup Budget Priorities Public Meeting, Richland, WA, June 7, 2017.
Pritikin, Trisha T. *The Hanford Plaintiffs: Voices from the Fight for Atomic Justice*. Lawrence: University Press of Kansas, 2020.
Proctor, Robert, and Londa L. Schiebinger. *Agnotology: The Making and Unmaking of Ignorance*. Stanford, CA: Stanford University Press, 2008.
Puar, Jasbir. *The Right to Maim: Debility, Capacity, Disability*. Durham, NC: Duke University Press, 2017.
"Radiation Safety Program Policies and Procedures." Los Angeles City College. https://www.lacitycollege.edu/Departments/Rad-Tech/documents/2020-LACC-Radiation-Safety-Program-(1).pdf.
Radiation Protection Criteria and Standards: Their Basis and Use; Summary-Analysis of the Hearings before the Special Subcommittee on Radiation of the Joint Committee on Atomic Energy. 86th Cong., 2nd Sess. May 24, 25, 26, 31 and June 1, 2, 3, 1960.
"Radioactive Corpse Left Decomposing." *Tennessean*. September 4, 1979.
"Radioactive Socks Found in Hanford Worker's Home." *Tri-City Herald*. October 1, 1998.
Radiology Support Devices, Inc. "The Fission-Product Phantom." https://rsdphantoms.com/product/the-fission-product-phantom/.
Raj, Ali. "In Marshall Islands, Radiation Threatens Tradition of Handing Down Stories by Song." *Los Angeles Times*. November 10, 2019.
"Rate of Melanoma Higher at Atom Lab." *Washington Post*. April 23, 1980.
RESRAD Workshop. Environmental Science Division, Argonne National Laboratory, 2018.

Rhodes, Richard. *The Making of the Atomic Bomb*. New York: Simon & Schuster, 1986.

Ridolfi, Callie. "Yakama Nation Exposure Scenario for Hanford Site Risk Assessment." Yakama Nation Environmental Restoration Waste Management Program. Union Gap, WA, 2007.

Roberts, Celia. "Drowning in a Sea of Estrogens: Sex Hormones, Sexual Reproduction and Sex." *Sexualities* 6, no. 2 (2003): 195–213.

Roberts, David. "Putting Fukushima into Perspective." *Wall Street Journal*. September 12, 2012.

Roff, Heather M. "Gendering a Warbot: Gender, Sex, and the Implications for the Future of War." *International Feminist Journal of Politics* 18, no. 1 (2016): 1–18.

Ropeik, David. "How the Unlucky Lucky Dragon Birthed an Era of Nuclear Fear." *Bulletin of Atomic Scientists*. February 28, 2018.

Rose, Nikolas S. *Politics of Life Itself: Biomedicine, Power, and Subjectivity in the Twenty-First Century*. Princeton, NJ: Princeton University Press, 2007.

Salmon, P. L., et al. "Alpha-Particle Doses to Cells of the Bone Remodeling Cycle from Alpha-Particle-Emitting Bone-Seekers: Indications of an Antiresorptive Effect of Actinides." *Radiation Research* 152, no. 6 (1999): S43–47.

Samatar, Sofia. *Tender: Stories*. Easthampton, MA: Small Beer Press, 2017.

Sanger, S. L. *Working on the Bomb: An Oral History of WWII Hanford*. Portland, OR: Portland State University Press, 1995.

Sarathy, Brinda, Vivien Hamilton, and Janet Farrell Brodie, eds. *Inevitably Toxic: Historical Perspectives on Contamination, Exposure, and Expertise*. Pittsburgh. PA: University of Pittsburgh Press, 2018.

Satariano, Adam. *The Robots Built to Clean up Our Nuclear Mess*. New York: Bloomberg, 2017.

Scarry, Elaine. *The Body in Pain: The Making and Unmaking of the World*. Oxford: Oxford University Press, 1985.

Scheck, Justin. "Bunnies Are in Deep Doo-Doo When They 'Go Nuclear' at Hanford." *Wall Street Journal*. December 23, 2010.

Scheper-Hughes, Nancy. "Dissection." In *A Companion to the Anthropology of the Body and Embodiment*, edited by Frances Mascia-Lees, 172–206. Oxford: Wiley-Blackwell, 2011.

Schieber, Caroline, and Christian Thézée. "Towards the Development of an ALARA Culture." *Proceedings of IRPA* 10 (2000): 1–4.

Schneer, Richard M. Letter to Linus Pauling. November 17, 1959. Oregon State University Special Collections and Archives Research Center, Ava Helen and Linus Pauling Papers, box 7.012, folder 12.17.

Schneider, Keith. "Washington Nuclear Plant Poses Risk for Indians." *New York Times*. September 3, 1990.

Schwab, Gabriele. *Radioactive Ghosts*. Minneapolis: University of Minnesota Press, 2020.

Schwartz, Stephen I. *Atomic Audit: The Costs and Consequences of U.S. Nuclear Weapons since 1940.* Washington, D.C.: Brookings Institution Press, 1998.

Scobie, William. "Plutonium Time-Bomb Threatens California." *London Observer.* May 18, 1980.

Shadaan, Reena, and M. Murphy. "EDC's as Industrial Chemicals and Settler Colonial Structures." *Catalyst: Feminism, Theory, Technoscience* 6, no. 1 (2020): 1–36.

Shapiro, Nicholas, Nasser Zakariya, and Jody Roberts. "A Wary Alliance: From Enumerating the Environment to Inviting Apprehension." *Engaging Science, Technology, and Society* 3 (2017): 575–602.

Shaw, Grazia Denig. Letter to Linus Pauling. May 24, 1961. Oregon State University Special Collections and Archives Research Center, Ava Helen and Linus Pauling Papers, box 7.014, folder 14.8.

Shinn, Lora. "As the DOE Abandons a Toxic Mess Threatening the Columbia River, the Yakama Nation Fights Back." Natural Resources Defense Council. September 19, 2019. https://www.nrdc.org/stories/doe-abandons-toxic-mess-threatening-columbia-river-yakama-nation-fights-back.

Silver, Ken. "The Energy Employees Occupational Illness Compensation Program Act: New Legislation to Compensate Affected Employees." *Workplace Health & Safety* 53, no. 6 (2005): 267–77.

Simon, Steven L., André Bouville, Charles E. Land, and Harold L. Beck. "Radiation Doses and Cancer Risks in the Marshall Islands Associated with Exposure to Radioactive Fallout from Bikini and Enewetak Nuclear Weapons Tests: Summary." *Health Physics* 99, no. 2 (2010): 105–23.

Sistrom, Joseph, Jerry Hopper, Sydney Boehm, Gene Barry, Lydia Clarke, Michael Moore, Nancy Gates, Lee Aaker, Leith Stevens, and Paramount Pictures Corporation. *The Atomic City.* Paramount, 1992.

Sontag, Susan. *Illness as Metaphor.* New York: Farrar, Straus and Giroux, 1978.

Stabin, M. G., et al. "Mathematical Models and Specific Absorbed Fractions of Photon Energy in the Nonpregnant Adult Female and at the End of Each Trimester of Pregnancy." ORNL/TM-12907. Oak Ridge, TN: Oak Ridge National Laboratory, 1995.

Stang, John. "Hanford Puzzled by Radioactive Garbage Bin." *Tri-City Herald.* October 2, 1998.

———. "Hanford Works to Trap Contaminated Bugs." *Tri-City Herald.* October 13, 1998.

———. "Hanford's Great Gator Escape Genuine." *Tri-City Herald.* May 26, 2002.

State and Tribal Government Working Group. *Closure for the Seventh Generation: A Report from the State and Tribal Government Working Group's Long-Term Stewardship Committee.* Denver, CO: National Conference of State Legislatures, 2017.

Stevenson, Lisa. *Life Beside Itself: Imagining Care in the Canadian Arctic.* Oakland: University of California Press, 2014.

Stifelman, Marc. "Letter to the Editor." *Risk Analysis* 23, no. 5 (2003); 859–60.

———. "Comments on the Yakama Nation Exposure Scenario for Hanford Risk Assessment." U.S. Environmental Protection Agency, Office of Environmental Assessment. January 3, 2008.

"Strontium May Be 60 Times Deadlier than the AEC Says." *I.F. Stone's Weekly* 7, no. 17 (May 4, 1959): 1–4.

"Summary of Remarks Prepared by Dr. Willard Libby, Commissioner United States Atomic Energy Commission for Delivery before the Swiss Academy of Medical Sciences Symposium on Radioactive Fallout." Lausanne, Switzerland. March 27, 1958. Oregon State University Special Collections and Archives Research Center, Ava Helen and Linus Pauling Papers, box 7.002, folder 2.11.

Szymendera, Scott D. *The Energy Employees Occupational Illness Compensation Act*. CRS Report: R46476. Washington, D.C.: Congressional Research Service, 2020.

Tabadatze, George, et al. "Re-evaluation of Am 241 Content in the USTUR Case 0102 Leg Phantom." *Health Physics* 104, no. 1 (2013): 9–14.

"Tainted Corpse Lay 4 Months in Mortuary." *Detroit Free Press*. September 4, 1979.

Takahashi, Tatsuya, Minouk Schoemaker, Klaus Trott, Steven Simon, Keisei Fujimori, Noriaki Nakashima, Akira Fukao, and Hiroshi Saito. "The Relationship of Thyroid Cancer with Radiation Exposure from Nuclear Weapon Testing in the Marshall Islands." *Journal of Epidemiology* 13, no. 2 (2003): 99–107.

Taylor, Frieda Yvonne. "History of the Lawrence Livermore National Laboratory Torso Phantom." ProQuest Dissertations Publishing, 1997.

Taylor, Lauriston. Letter to Strauss, November 10, 1948. Lauriston Sale Taylor Papers, box 31, file: "NCRP-1948."

"ThermoFisher Scientific." *Radwaste Solutions* 25, no. 1 (2018): 49.

Thompson, Dorothy. "Radioactivity and the Human Race." *Ladies Home Journal*. September 11, 1956.

Thompson, R. C. *Pacific Northwest National Laboratory Annual Report for 1971 to the USAEC Division of Biology and Medicine*. AEC Research and Development Report. BNWL-1650-PT1. Richland, WA, September 1972.

United Nations Scientific Committee on the Effects of Atomic Radiation. *Sources and Effects of Ionizing Radiation: UNSCEAR 2000 Report to the General Assembly*. New York: United Nations, 2000.

U.S. Department of Energy. *Closing the Circle on the Splitting of the Atom: The Environmental Legacy of Nuclear Weapons Production in the United States and What the Department of Energy Is Doing about It*. Washington, D.C.: U.S. Department of Energy, Office of Environmental Management; U.S. GPO, distributor, 1995.

———. *Estimating the Cold War Mortgage: The Baseline Environmental Management Report*. Washington, D.C.: U.S. Department of Energy, Office of Environmental Management, 1995.

———. *Linking Legacies: Connecting the Cold War Nuclear Weapons Production Processes to Their Environmental Consequences.* Washington, D.C.: U.S. Department of Energy, Office of Environmental Management, 1997.

———. "The Department of Energy Laboratory Accreditation Program for Radiobioassay." DOE-STD-1112-98. Washington, D.C., 1998.

———. "Fall 1998 200 East Area Biological Vector Contamination Report." HNF-3628. Richland, WA, 1999.

———. "Final Hanford Comprehensive Land-Use Plan Environmental Impact Statement." DOE/EIS-0222-F. Washington, D.C., 1999.

———. "Use of Risk-Based End States." DOE Policy 455.1. Office of Environmental Management, Washington, D.C., 2003.

———. "Supplemental Analysis: Hanford Comprehensive Land-Use Plan Environmental Impact Statement." DOE/EIS-0222-SA-01. Richland, WA, 2008.

———. "Radiation Protection of the Public and the Environment." DOE O 458.1. Office of Environment, Health, Safety, and Security. Washington, D.C., 2011.

———. "Hanford Long-Term Stewardship Program Plan." DOE/RL-2010-35, Rev. 1. Richland, WA, April 2012.

———. *DOE Handbook: Radiological Worker Training.* DOE-HDBK-1130-2008. Washington, D.C.: DOE, 2013.

———. "Hanford Site Cleanup Completion Framework." DOE/RL-2009-10, Rev. 1. Richland, WA, 2013.

———. "Time of Compliance for Disposal of Low-Level Radioactive Waste." Office of Environmental Health, Safety, and Security. August 11, 2014. https://www.energy.gov/ehss/downloads/time-compliance-disposal-low-level-radioactive-waste.

———. "Supplemental Analysis of the Hanford Comprehensive Land-Use Plan Environmental Impact Statement." DOE/EIS-0222-SA-02. Richland, WA, 2015.

———. "Office of Enterprise Assessments Follow-Up Assessment of Progress on Actions Taken to Address Tank Vapor Concerns at the Hanford Site." Office of Worker Health and Safety Assessments, Office of Environment, Safety, and Health Assessments. January 2017.

———. "Office of Enterprise Assessments Follow-Up Assessment of Progress on Actions Taken to Address Tank Vapor Concerns at the Hanford Site." Office of Worker Health and Safety Assessments, Office of Environment, Safety, and Health Assessments. February 2018.

———. "2019 Hanford Lifecycle Scope, Schedule, and Cost Report." DOE/RL-2018-45. Richland, WA, 2019.

———. "Hanford Annual Site Environmental Report for Calendar Year 2020." DOE/RL-2021-15. Richland, WA, 2021.

———. "Hanford Site Groundwater Monitoring Report for 2020." DOE/RL-2020-60. Richland, WA, 2021.

———. "Assessment of Department of Energy's Interpretation of the Definition of High-Level Radioactive Waste." *Federal Register* 86, no. 242 (December 21, 2021).

———. "About Hanford Cleanup." https://www.hanford.gov/page.cfm/About HanfordCleanup.

———. "Hanford Advisory Board." https://www.hanford.gov/page.cfm/hab.

———. "Phantom Library." Office of Environment, Health, Safety, and Security https://www.energy.gov/ehss/phantom-library.

U.S. Environmental Protection Agency. "Memorandum: Land Use in the CERCLA Remedy Selection Process." Office of Solid Waste and Emergency Response. OSWER Directive No. 9355.7-04. Washington, D.C., May 25, 1995.

———. "Residual Risk: Report to Congress." Office of Air Quality Planning and Standards. EPA-453/R-99-001. Research Triangle Park, NC. March 1999.

———. *Rules of Thumb for Superfund Remedy Selection.* EPA 540-R-97-013. OSWER 9355.0-69. Washington, D.C.: Office of Solid Waste and Emergency Response, 1997.

———. "Comprehensive Five-Year Review Guidance." Office of Solid Waste and Emergency Response. OSWER No. 9355.7-03B-P. Washington, D.C., July 17, 2001.

———. *Staff Paper: Risk Assessment Principles and Practices.* Washington, D.C.: U.S. Environmental Protection Agency, Office of the Science Advisor, 2004.

———. "Public Health and Environmental Radiation Protection Standards for Yucca Mountain, Nevada." 40 CFR Part 197. *Federal Register* 73, no. 200 (October 15, 2008).

———. Superfund Site Remediation Program: Radiation Risk Assessment Training. Phoenix, AZ, March 18, 2018.

———. *Guidelines for Human Exposure Assessment.* EPA/100/B-19/001. Washington, D.C.: Risk Assessment Forum, 2019.

———. "Conducting a Human Health Risk Assessment." n.d. https://www.epa.gov/risk/conducting-human-health-risk-assessment#tab-5.

———. "Consumption by Tribes of Plants and Animals Not Accounted for in EPA Superfund Risk Assessment Methodology." Office of Superfund Remediation and Technology Innovation (OSRTI). n.d. https://clu-in.org/conf/tio/plantcbyt_060320/.

———. "Radionuclide Basics: Plutonium." n.d. https://www.epa.gov/radiation/radionuclide-basics-plutonium#plutoniumhealth.

———. "Superfund: Institutional Controls." n.d. https://www.epa.gov/superfund/superfund-institutional-controls.

———. "Superfund Remedy Overview, Key Principles and Guidance." n.d. https://www.epa.gov/superfund/superfund-remedy-overview-key-principles-and-guidance.

U.S. Federal Civil Defense Administration. *Survival under Atomic Attack.* Document 130. Washington, D.C.: Executive Office of the President, National Security Resources Board, Civil Defense Office, 1951.
———. *Facts about Fallout.* Washington, D.C.: U.S. GPO, 1955.
U.S. National Bureau of Standards. *Permissible Dose from External Sources of Ionizing Radiation: Recommendations of the National Committee on Radiation Protection.* Washington, D.C.: U.S. Department of Commerce, National Bureau of Standards; U.S. GPO, 1954.
U.S. Nuclear Regulatory Commission, Issuing Body. *The Nuclear Waste Policy Act of 1982.* Washington, D.C.: U.S. Nuclear Regulatory Commission, 1983.
———. "Reassessment of the NRC's Dollar per Person-Rem Conversion Factor Policy." Federal Register 80, no. 172 (September 4, 2015): 53585.
———. "Occupational ALARA and Planning Controls." In *NRC Inspection Manual.* March 2, 2016. https://www.nrc.govdocs/ML1534/ML15344 A278.pdf.
U.S. Nuclear Regulatory Commission. "Radiation Basics." https://www.nrc.gov/about-nrc/radiation/health-effects/radiation-basics.html#.
U.S. Transuranium and Uranium Registries. "Hanford Autopsy Study." https://ustur.wsu.edu/history/hanford-study/.
———. "National Human Radiobiology Tissue Repository." https://ustur.wsu.edu/nhrtr/.
———. "Prospective Collaborators." https://ustur.wsu.edu/collaborators/.
Vedantham, S., and A. Karellas. "SU-E-I-61: Phantom Design for Phase Contrast Breast Imaging." *Medical Physics* 39, no. 6 (2012): 3638–39.
Veolia. "Veolia Highlights World-Class Capabilities as Global Leaders in Radioactive Waste Industry Gather at WMS 2018." 2018. https://www.nuclearsolutions.veolia.com/en/news/veolia-highlights-world-class-capabilities-global-leaders-radioactive-waste-industry-gather.
Viscusi, W. Kip. *Fatal Tradeoffs: Public and Private Responsibilities for Risk.* Oxford: Oxford University Press, 1996.
Voelz, G. L., J. N. P. Lawrence, and E. R. Johnson. "Fifty Years of Plutonium Exposure to the Manhattan Project Plutonium Workers: An Update." *Health Physics* 73, no. 4 (1997): 611–19.
Wald, Matthew. "Even Rabbit Droppings Count in Nuclear Cleanup." *New York Times.* October 14, 2009.
Waldby, Catherine. *The Visible Human Project: Informatic Bodies and Posthuman Medicine.* New York: Routledge, 2000.
Walker, J. Samuel. *Permissible Dose: A History of Radiation Protection in the Twentieth Century.* Berkeley: University of California Press, 2000.
Wallace, L. A. "Human Exposure to Environmental Pollutants: A Decade of Experience." *Clinical and Experimental Allergy* 25 (1995): 4–9.
Wanapum Heritage Center. "Repository." https://wanapum.org/about/repository.

Warrick, Joby. "Panel Links Illness to Nuclear Work; Former Energy Dept. Employees May Get Compensation for Exposure." *Washington Post.* January 30, 2000.

Washington River Protection Solutions. "ALARA Work Planning." Document TFC-ESHQ-RP_RWP-C-03. 2017. https://hanfordvapors.com/wp-content/uploads/2018/08/TFC-ESHQ-RP_RWP-C-03-ALARA-Work-Planning.pdf.

Washington State Attorney General. "Hanford." https://www.atg.wa.gov/hanford.

Washington State Department of Health. "Hanford Environmental Radiation Oversight Program: 2017 Data Summary Report." Publication 320-124. Office of Radiation Protection. Richland, WA, 2019.

———. "Hanford and Public Health." 2018. https://doh.wa.gov/community-and-environment/radiation/radiation-topics/hanford-and-public-health.

Waste Management Symposia. "Step Up to the Golf Simulator." *Insight* 44, no. 1 (2018): 1–8.

———. "Why Attend?" 2018. https://www.wmsym.org/register/why-attend/.

———. "Why Exhibit?" 2018. https://www.wmsym.org/exhibitors/why-exhibit/.

Watson, C. R., and R. O. McClellan. "In Vivo Thermoluminescence Dosimetry of Gamma Rays from Ingested Cs-137 in Sheep." In *Luminescence Dosimetry*, edited by Frank H. Attix, 393–401. Springfield, VA: U.S. Atomic Energy Commission, Division of Technical Information, National Bureau of Standards, U.S. Department of Commerce, 1967.

Weber, Max. *From Max Weber: Essays in Sociology.* London: Routledge, 1952.

Weinstein, B. L. Letter to Linus Pauling. October 12, 1959. Oregon State University Special Collections and Archives Research Center, Ava Helen and Linus Pauling Papers, box 7.012, folder 12.17.

Welshons, Wade V., Kristina A. Thayer, Barbara M. Judy, Julia A. Taylor, Edward M. Curran, and Frederick S. vom Saal. "Large Effects from Small Exposures: I. Mechanisms for Endocrine-Disrupting Chemicals with Estrogenic Activity." *Environmental Health Perspectives* 111, no. 8 (2003): 994–1006.

Welsome, Eileen. *The Plutonium Files: America's Secret Medical Experiments in the Cold War.* New York: Dial Press, 1999.

"Whistleblower Retaliation at the Hanford Nuclear Site: Hearing before the Subcommittee on Financial and Contracting Oversight of the Committee on Homeland Security and Governmental Affairs, U.S. Senate. 113th Cong., 2nd Sess. March 11, 2014.

White, D. R. "Tissue Substitutes in Experimental Radiation Physics." *Medical Physics* 5, no. 6 (1978): 467–79.

Whittemore, Gilbert Franklin. "The National Committee on Radiation Protection, 1928–1960: From Professional Guidelines to Government Regulation." Harvard University. Pro Quest Dissertations Publishing, 1986.

Wilson, Richard. "Risks Caused by Low Levels of Pollution." *Yale Journal of Biology and Medicine* 51 (1978): 37–51.

Xin, Ling. "Bulletin of the Atomic Scientists Moves Doomsday Clock 2 Minutes Closer to Midnight." *Science*. January 22, 2015.

Xu, Xie George, and K. F. Eckerman. *Handbook of Anatomical Models for Radiation Dosimetry*. Boca Raton, FL: CRC Press/Taylor & Francis Group, 2010.

Yakama Nation. "Yakama Nation Treaty of 1855." https://www.yakama.com/about/treaty/.

Young, Monica. "Meet Valkyrie, NASA's Space Robot." *Sky and Telescope*. May 17, 2017.

Yu, C., A.J. Zielen, J.-J. Cheng, D.J. LePoire, E. Gnanapragasam, S. Kamboj, J. Arnish, A. Wallo III, W.A. Williams, and H. Peterson. *User's Manual for RESRAD Version 6*. Argonne, IL: Argonne National Laboratory, 2001.

Yu, Charley, Sunita Kamboj, Cheng Wang, and Jing-Jy Cheng. *Data Collection Handbook to Support Modeling Impacts of Radioactive Material in Soil and Building Structures*. Argonne, IL: Argonne National Laboratory, 2015.

Index

acceptable risk, 15; acceptance or lack of acceptance of, 83–86; ALARA (as low as reasonably achievable) principle, 96–98; autopsy studies and, 53–61; EEOICPA's compensation process, 101–6; In Vivo Radiobioassay and Research Facility (IVRRF), 42–53, 42*fig*, 45*fig*, 46*fig*, 48*fig*, 50*fig*, 52*fig*; JCAE, *Radiation Protection Criteria and Standards: Their Basis and Use*, 94–96; *Radiological Worker Training* handbook, DOE, 97, 98–99; Reference Person (Standard Man, Reference Man), 34–40, 46–47, 49, 51, 93, 147n42, 149–50n61, 149n57; regulatory frameworks, development of, 24–26, 143n70; statistical lives in, 51, 59–60; workers' embodied knowledge of exposure, 99–101. *See also* reasonable harm; risk and risk assessments
acute radiation disease, 70
AEC. *See* Atomic Energy Commission (AEC)
Aiken, South Carolina, 11
ALARA (as low as reasonably achievable) principle, 96–98
Alvarez, Bob, 11

Amarillo, Texas, 11
americium-241, 43; autopsy studies and, 41, 56–61; Stuart R. Gunn, Case 0102, 41, 61–66, 64*fig*
Argonne National Laboratory: RESRAD software and assessment of radiation exposure, 28–40, 32*fig*, 40*fig*
as low as reasonably achievable (ALARA) principle, 96–98
atomic bomb. *See* nuclear weapons
atomic bomb disease, 70
The Atomic City (film), 61
Atomic Energy Commission (AEC), 165n82; Health and Safety Laboratory, 85; National Academy of Sciences, genetics committee, 72–77; National Plutonium Registry, 54; nuclear fallout, study of and debate about, 67–72, 83–85
autopsy studies, 53–61

background radiation, 43, 70, 79
Bair, William (Bill), 175–76n41
Beadle, George, 74
BEAR Report. *See* Biological Effects of Atomic Radiation (BEAR) Report
Benjamin, Ruha, 129–30

207

Biological Effects of Atomic Radiation (BEAR) Report, 70–71, 83–86
biotic vectors of contamination, 115–19
body burden: ALARA (as low as reasonably achievable) principle, 96–98; body as lived practice and regulatory device, 16; DOE listening tour in Richland, 87–89; EEOICPA (Energy Employees Occupational Illness Compensation Program Act) (2000), 89–90; EEOICPA's compensation process, 101–6; JCAE, *Radiation Protection Criteria and Standards: Their Basis and Use*, 94–96; radiation effects, differences in radiation types and body tissues, 95–96; *Radiological Worker Training* handbook, DOE, 97, 98–99; Special Exposure Cohort (SEC) status, 105–6; workers' embodied knowledge of exposure, 99–101. *See also* reasonable harm; risk and risk assessments
Bohr, Niels, 11
Bombproof (FCDA film), 81
bone tissue, 96; autopsy studies, 53, 55, 56, 61–62, 65, 69; radiation exposure, 87, 96, 104, 176n41; phantoms and, 41–42, 44–45, 51, 158n72, 159n86; Reference Man standards, 36
Boyd, William, 25, 145–46n20
B reactor, Hanford Nuclear Reservation, 60, 91*fig*, 116, 120–26
breast tissue: phantom testing and breast tissue, 47–49, 153–54n20, 154n27; Reference Person (Standard Man, Reference Man), risk assessments and, 35–40, 149n57
Bronk, Detlev, 70–71
budget for cleanup, 7–8, 19–20, 20*fig*, 29, 112
Buske, Norm, 115–16
Butler, Judith, 66, 139n27

Canada, Deep Geological Repository, 29
cancer, 3–4; author's personal context, 5–6, 62, 63; Reference Person, risk assessments and, 35–40; risk as concept in U.S. public health regulations, 15, 24–26, 98–99, 104–6, 137n20, 143n70, 151–52n73, 171n65. *See also* body burden; reasonable harm; Reference Person (Standard Man, Reference Man); risk and risk assessments
Case 0102: phantoms, 41–42, 42*fig*, 51–52, 53, 158n72, 159n82, 159n85, 159n86; Stuart R. Gunn, 61–66, 64*fig*

Castle-Bravo detonation, 69–70
Central Plateau, Hanford Nuclear Reservation, 19, 110
Centre d'etude sur l'Evaluation de la Protection dans le domaine Nucleaire (roughly, Center for the Study of Nuclear Protection), 97
cesium-137, 19, 30
Chen, Mel, 145n14
Chernobyl, 71
children: civil defense drills for, 14, 80; EEOICPA lack of remedies for workers' families, 106; nuclear fallout, study of effects of, 69, 76, 77, 78, 79, 164n59; Reference Person (Standard Man, Reference Man), risk assessments and, 35–40, 149n61
cleanup-as-containment, 3, 13, 114, 118. *See also* environmental cleanup efforts
Clinton (Bill) administration, DOE secretary Bill Richardson, 89–90
Closing the Circle on the Splitting of the Atom: The Environmental Legacy of Nuclear Weapons Production and What the Department of Energy Is Doing about It, 12
CLUP (comprehensive land use plan), DOE, 109–15
Cold War: economization of life and, 58–60; end of, lack of planning for, 11–12; failures of nuclear imagination during, 28; medical experimentation in, 58; nuclear pedagogy, contradictions in, 9–10; relationship between injury and national security, 71–72
Collier's, 80
Columbia River, 2, 90–94, 110; Hanford Reach, environmental monitoring on, 107–9, 109*fig*
Comprehensive Environmental Response, Compensation, and Liability Act (CERCLA), 111–15, 174n31
comprehensive land use plan (CLUP), DOE, 109–15
Confederated Tribes of the Umatilla Indian Reservation (CTUIR), 38–39, 136n9, 150n63, 150n64, 151n70, 173n10
contamination, 144n76; autopsy studies and, 56–61; questions about, 4; RESRAD software and assessment of radiation exposure, 28–40, 32*fig*. *See also* environmental cleanup efforts; risk and risk assessments
cost-benefit analysis: ALARA (as low as reasonably achievable) principle, 96–98; CERCLA (Comprehensive Environmental

Response, Compensation, and Liability Act) and, 112; in environmental policy, 25–26; ICRP Publication 103, exposure risk assessments and, 37; JCAE, *Radiation Protection Criteria and Standards: Their Basis and Use*, 94–96; National Academy of Sciences, genetics committee negotiations, 72–77; relationship between injury and progress, 71–72. *See also* reasonable harm; risk and risk assessments

Daigo Fukuryu Maru (Lucky Dragon No. 5), 69–70
Deep Geological Repository (Canada), 29
de-identification, 56–60; Stuart R. Gunn, Case 0102, 61–66, 64*fig*
Demerec, Milislav, 76–77
detrimentally affected person, 75–76
dosimetry: autopsy studies and risk assessment, 56–61; EEOICPA's compensation process, 101–6; 109*fig*; IVRRF (In Vivo Radiobioassay and Research Facility), 16, 42–53, 42*fig*, 45*fig*, 46*fig*, 48*fig*, 50*fig*, 52*fig*; phantom testing and breast tissue, 47–49, 148n48; Lawrence Livermore National Laboratory (LLNL) Realistic Torso Phantom, 46–47, 46*fig*, 48*fig*; Reference Person (Standard Man, Reference Man), 34–40, 46–47, 49, 51, 93, 147n42, 149–50n61, 149n57; USTUR, 54–58, 60–61, 157n66
duck and cover drills, 14

economic issues: aging waste facilities and budget constraints, 7–8, 19–20, 20*fig*, 29, 112; Cold War and the economization of life, 58–60; nuclear weapons production, scale of, 11; statistical person and, 59–60
Edwards, Paul, 9
EEOICPA. *See* Energy Employees Occupational Illness Compensation Program Act (2000) (EEOICPA)
Eisenbud, Merril, 85
embodying the insensible, 79–80, 100, 123–124
emotions, management of, 10, 14, 80–83
Energy Employees Occupational Illness Compensation Program Act (2000) (EEOICPA), 90, 94; compensation process, 101–6; Special Exposure Cohort (SEC) status, 105–6
environmental cleanup efforts: budget constraints for, 7–8, 19–20, 20*fig*, 29, 106, 112;

DOE Office of Environmental Restoration and Waste Management, 11–12; National Defense Authorization Act (1995), 12–13; questions about, 3–4, 6–9, 16; remediation, use of term, 3, 126; RESRAD (REsidual RADiation) software, 28–40, 32*fig*; robotics, use in waste management, 17–18, 20–21, 26–28; Superfund National Priority List, 15. *See also* Hanford Nuclear Reservation
environmental justice, 8–9, 37–39, 113, 119, 151n73, 174n31
environmental monitoring: Hanford activities, 107–108, 115; nuclear fallout, study of and debate about, 68–72; technology for, 24–25
environmental policy, U.S., 3, 24–26, 111–115
environmental sensitivity, 21–24, 38
Estimating the Cold War Mortgage: The 1995 Baseline Environmental Management Report, 12
Experimental Animal Farm, Hanford, 43, 116, 175n41
Exxon Valdez, 2

Facts about Fallout (FCDA, 1955), 80
Failla, Gioacchino, 73
fallout shelters, 14, 82
fear, emotional management and, 10, 14, 80–83
Federal Civil Defense Administration (FCDA), 80–83
Fission-Product Phantom, 50–51, 50*fig*
Food and Drug Administration (FDA), acceptable risk for toxic residues, 24–25
fruit fly incident, 116–18
Fukushima, Japan, 2

gender: female phantoms, lack of, 49; phantom testing and breast tissue, 47–49; Reference Person (Standard Man, Reference Man), risk assessments and, 35–40, 163n44
genetics: nuclear fallout, study of and debate about, 71–77; regulation of environmental contamination due to fallout, 77–86
George, Marlene, 113
Godzilla, 70, 155n39, 162n21
Gordon, Avery, 65, 66
groundwater contamination: budgetary constraints and cleanup priorities, 19–20; at Hanford Nuclear Reservation, 1–2, 3, 114,

groundwater contamination (*continued*)
129; RESRAD software and assessment of radiation exposure, 28–40, 32*fig*
grout, 127, 128*fig*
Grumbly, Thomas, 12
Gunn, Alan, 61–66, 64*fig*
Gunn, Stuart R., 61–66, 64*fig*

Hanford Advisory Board (HAB), 2, 28, 33, 38, 43, 114, 115, 127, 128*fig*, 136n9, 138n25; aging waste facilities and budget constraints, 18–20, 20*fig*
Hanford Autopsy Study, 53
Hanford Challenge, 90–94
Hanford Nuclear Reservation, 1–9, 15–16, 116*fig*, 135n1; B reactor, 91*fig*, 120–26; CERCLA (Comprehensive Environmental Response, Compensation, and Liability Act) and, 111–15; CLUP (comprehensive land use plan), DOE and, 109–15; Experimental Animal Farm, 43, 116, 175–76n41; impossibilities of cleanup, 7–8, 16, 129–130, 138–39n25; Indigenous peoples treaty rights, 38–39, 110–11, 150n64, 151n68, 174n31, 179n21; Internal Dosimetry Program, 42–43; invisible boundaries of, 107–9, 109*fig*, 115–19; IVRRF (In Vivo Radiobioassay and Research Facility), 42–53, 42*fig*, 45*fig*, 46*fig*, 48*fig*, 50*fig*, 52*fig*; storage tunnel collapse (2018), 18–19; waste reclassification, 127–30; worker exposure, 23, 27; 87–94, 99–101, 106, 123–26
Hanford Reach, 90–91, 107–9, 109*fig*
Hanford Site Cleanup Completion Framework, DOE, 113–15
Hanford Long-Term Stewardship (LTS) program, 113–15
Haraway, Donna, 94
Hartman, Saidiya, 143–44n75
hazardous materials: aging waste facilities and budget constraints, 19–20, 20*fig*; questions about, 4; RESRAD software and assessment of radiation exposure, 28–40, 32*fig*; robotics and waste management, 17–18, 20–21, 26–28. *See also* environmental cleanup efforts; risk and risk assessments
health issues: EEOICPA's compensation process, 101–6; physical responses to toxic exposures, 20–24
Healy, Jack, 95

Hecht, Gabrielle, 13–14
Heinzerling, Lisa, 51, 59, 60, 158–59n74
Hepler-Smith, Evan, 8, 140n31
Hiroshima, 70, 104
Hurley, Jessica, 10–11
Hutt, Peter, 25

Idaho Falls, Idaho, 11
insect vectors of contamination, 116–19
institutional controls, 3, 108–15, 137n15, 139n26, 172n5
Intercalibration Committee for Low-Photon Energy Measurements, DOE, 46
Internal Dosimetry Program, Hanford, 42–43
International Commission on Radiation Protection (ICRP), 149–50n61, 149n57; Publication 23, 34–40; Publication 103, 36–37; Publication 130, 36, 148n54
In Vivo Radiobioassay and Research Facility (IVRRF), 41–53, 42*fig*, 45*fig*, 46*fig*, 48*fig*, 50*fig*, 52*fig*

Jain, S. Lochlann, 6
Japan: atomic bomb disease, 70; *Daigo Fukuryu Maru* (*Lucky Dragon No. 5*), 69–70; Hiroshima and Nagasaki, 70, 104
JCAE. *See* Joint Committee on Atomic Energy (JCAE)
Jefferson County, Colorado, 11
Jim, Russell, 151–52n73, 174n31
Joint Committee on Atomic Energy (JCAE), 84–85, 94–96
Jose, Donald, 97–98

Kim Jong-un, 54
Kornberg, H. A., 175–76n41
Krupar, Shiloh, 104, 165n69

Laboratory Accreditation Program, DOE (DOELAP), 44, 49, 154n20
Ladies Home Journal, 83
land use, DOE's planning for, 109–15
Lawrence Livermore National Laboratory (LLNL) Realistic Torso Phantom, 46–47, 46*fig*, 48*fig*, 49, 50
legal issues: CERCLA (Comprehensive Environmental Response, Compensation, and Liability Act), 111–15; definitions of exposure and negligence claims, 97–98; EEOICPA's compensation process, 101–6; regulatory frameworks for acceptable risk,

24–26; standard of care, 169n37; waste
    reclassification, 127–30
Let's Face It (FCDA film), 81
Libby, Willard, 74, 85
life expectancy, statistical person and, 59–60
lifetime exposure, permissible dose and, 74
*Linking Legacies: Connecting the Cold War
    Nuclear Weapons Production Processes to
    their Environmental Consequences*, 12
Little, C. C., 73
Long-Term Stewardship (LTS) program,
    Hanford, 113–15
Los Alamos, New Mexico, 11
LTS (Long-Term Stewardship) program,
    Hanford, 113–15
lungs: autopsy studies, 53, 56; cancer of
    workers, 87, 104; dosimeter use and, 103;
    phantoms and, 41, 43, 51; Reference Man
    (Standard Man, Reference Person) and,
    35, 36; testing at In Vivo Radiobioassay
    and Research Facility (IVRRF), 44–47,
    45*fig*, 48*fig*. *See also* body burden

Manhattan Project, 11, 58, 121, 135n1
Manhattan Project National Historical Park, 121
Marshall Islands, 69–70
Masco, Joseph, 14, 15, 58, 80
medical experimentation in Cold War, 58
Michaels, David, 101
Morgan, Edward P., 83
Morgan, Jennifer, 139–40n28
Muller, Herman, 73, 75, 76
Murphy, M., 9, 58–59, 130
mutation. *See* genetics
mythology, Plutonium, 121–26

Nagasaki, 70, 104
NAS. *See* National Academy of Sciences
    (NAS)
NASA (National Aeronautics and Space
    Administration), 17–18
Nash, Linda, 15, 25–26
National Academy of Sciences (NAS): genetics
    committee, negotiations of, 72–77; nuclear
    fallout, study of and debate about, 70–71,
    83–86
National Bureau of Standards Handbook, 74
National Defense Authorization Act (1995), 12
National Environmental Policy Act (NEPA),
    112, 174n28
National Human Radiobiology Tissue Repos-
    itory (NHRTR), 54–61

National Plutonium Registry, 54
National Research Council (NRC), 25, 118–19
Native Americans, 151–52n73, 151n70, 174n31,
    179n21; CERCLA and, 113; CLUP (com-
    prehensive land use plan) DOE, 110–15;
    nuclear fuel cycle on tribal lands, 38;
    treaty rights of, 38–39, 110–11, 150n64,
    151n68, 174n31, 179n21
NCRP. *See* U.S. National Committee on
    Radiation Protection (NCRP)
Neel, James, 73, 76–77
NEPA (National Environmental Policy Act),
    112, 174n28
Newell, R. R., 85
New York Times, 68
Nez Perce Tribe, 136n9, 150n64
NHRTR (National Human Radiobiology
    Tissue Repository), 54–61
Niles, Ken, 111
North Korea, 54
No Trespassing signs, 2, 3, 109*fig*, 130
NRC (National Research Council), 25, 118–19
nuclear exceptionalism, 14
nuclear fallout: aboveground nuclear testing,
    24, 69–70; civil defense agencies, public
    safety campaigns of, 10, 14, 80–83; regula-
    tion of environmental contamination due
    to fallout, 77–86; study of and debate
    about, 15, 53–54, 67–77
nuclear imagination, failures of, 28
nuclear mundane, 11
Nuclear Safety Division, Oregon Department
    of Energy, 111, 128
nuclear sublime, 10–11
nuclear waste: aging waste facilities and
    budget constraints, 19–20, 20*fig*; Hanford
    Nuclear Reservation, 1–9, 13, 91–94, 91*fig*,
    107–9, 120–30, 135n1, 138–39n25;
    RESRAD software and assessment of
    radiation exposure, 28–40, 32*fig*; robotics
    and waste management, 17–21, 26–28;
    waste reclassification, 127–30
nuclear weapons: aboveground nuclear test-
    ing, 24, 69–70; Castle-Bravo detonation,
    69–70; civil defense agencies, public
    safety campaigns of, 10, 80–83; end of
    Cold War, lack of planning for, 11–12;
    environmental legacy as untold story, 11,
    25; Hiroshima, 70, 104; industrial produc-
    tion, scale of, 1, 11; Manhattan Project
    National Historical Park, 121; mushroom
    cloud as abstraction, 9–11; Nagasaki, 70,

nuclear weapons (*continued*)
104; North Korea and, 54; nuclear pedagogy, contradictions in, 9–10; Richland High School mascot, 88, 88*fig*, 89*fig*. *See also* nuclear fallout

Oak Ridge, Tennessee, 11
occupational illness. *See* body burden
OCD. *See* U.S. Office of Civil Defense (OCD)
Office of Environmental Restoration and Waste Management, DOE, 11–12, 141n48
O'Leary, Hazel, 11, 12, 141n49
Oppenheimer, Robert, 39
Oregon Department of Energy, Nuclear Safety Division, 111, 128
ovarian tissue, 35–40

Parker, Herbert, 96
Parr, Joy, 100
Pauling, Ava Helen, 68
Pauling, Linus, 68, 78–79
permissible dose: as a conceptual framework, 14; debate over use of term, 73–75, 77; EEOICPA's compensation process, 94, 101–6; Hanford Autopsy Study and, 53; in *National Bureau of Standards Handbook*, 74; nuclear workers, higher permissible doses for, 84–85, 94, 98; *Radiological Worker Training* handbook, DOE, 97, 98–99; regulatory policy and, 78, 84–85, 165–66n86; workers' embodied knowledge of exposure, 16, 99–101. *See also* risk and risk assessments
Petersen, Gary, 175–76n41
Peterson, Val, 80, 81
phantoms: Case 0102, phantom arm, 41–42, 42*fig*, 51–52, 53; defined, 42; Fission-Product Phantom, 50–51, 50*fig*; Lawrence Livermore National Laboratory (LLNL) Realistic Torso Phantom 46–47, 46*fig*, 48*fig*; lung and liver phantoms, 44, 45*fig*, phantom testing and breast tissue, 47–49; Stuart R. Gunn, Case 0102, 61–66, 64*fig*; testing at In Vivo Radiobioassay and Research Facility (IVRRF), 42–53, 42*fig*, 45*fig*, 46*fig*, 48*fig*, 50*fig*, 52*fig*; thyroid phantom, 50–51, 52*fig*. *See also* Reference Person (Standard Man, Reference Man)
physical responses to toxic exposures, 20–24
plant vectors of contamination, 115–19
Playboy, 68

plutonium, 135n1; autopsy studies and, 56–61; Hanford Autopsy Study, 53; Hanford Nuclear Reservation, 1–9
Plutonium, mythology of, 121–26
pregnancy, Reference Person and risk assessments, 35–40
Publication 23 (ICRP), 34–40
Publication 103 (ICRP), 36–37
Publication 130 (ICRP), 36, 148n54
public health regulations: Reference Person (Standard Man, Reference Man), 34–40, 46–47, 49, 51, 93, 147n42, 149–50n61, 149n57; risk as a concept in, 15, 24–26, 98–99, 104–6, 137n20, 143n70, 151–52n73, 171n65

quality-adjusted life years (QALYs), 60

Radiation Protection Criteria and Standards: Their Basis and Use, 94–96
radiation sickness, 81, 176n41
radioactive waste. *See* nuclear waste
Radiological and Environmental Sciences Laboratory (RESL), 44
radiological dose assessment, RESRAD software, 28–40, 32*fig*. *See also* dosimetry
Radiological Safety Officer Magazine, 97–98
Radiological Worker Training handbook, DOE, 97, 98–99
RADS (reactive airways dysfunction syndrome), 93
reasonable harm: National Academy of Sciences, genetics committee negotiations, 72–77; nuclear fallout, study of and debate about, 68–72; Reference Person (Standard Man, Reference Man), 34–40, 46–47, 49, 51, 93, 147n42, 149–50n61, 149n57; regulation of environmental contamination due to fallout, 77–86; relationship between injury and progress, 71–72; responses to radiation exposure risk, 83–86; within the definition of cleanup, 4, 13, 16, 130. *See also* body burden
receptor, 16, 33–38, 128
Reference Person (Standard Man, Reference Man), 34–40, 46–47, 49, 51, 93, 147n42, 149–50n61, 149n57. *See also* phantoms; receptor
regulatory efforts: impossibilities of, 7–9, 118–19; Reference Person (Standard Man, Reference Man), 34–40, 46–47, 49, 51, 93, 147n42, 149–50n61, 149n57; regulation of

INDEX 213

environmental contamination due to fallout, 77–86; risk as a concept in U.S. public health regulations, 15, 24–26, 98–99, 104–6, 137n20, 143n70, 151–52n73, 171n65. *See also* specific agency names
remediation: budget constraints for, 19–20, 20*fig*; CERCLA (Comprehensive Environmental Response, Compensation, and Liability Act), 111–15; CLUP (comprehensive land use plan), DOE, 109–15; Hanford Nuclear Reservation, 2–9, 15, 138–39n25; on Hanford Reach, 107–9, 109*fig*; RESRAD software and assessment of radiation exposure, 28–40, 32*fig*; as a social product, 3–9, 129; U.S. Department of Energy mandates for, 12–13; waste reclassification, 127–30. *See also* environmental cleanup efforts
RESL (Radiological and Environmental Sciences Laboratory), 44
RESRAD (RESidual RADiation) software, 28–40, 32*fig*; model as imperfect representation, 31–33, 51
Richardson, Bill, 89–90
Richland, Washington, 11; DOE listening tour for nuclear workers, 87–89; high school mascot, 88, 88*fig*, 89*fig*; radioactive flies in, 117–18; tumbleweeds, contamination of, 115, 116*fig*
risk and risk assessments: autopsy studies and, 53–61; EEOICPA's compensation process, 101–6; ICRP Publication 103, exposure risk assessments and, 37; In Vivo Radiobioassay and Research Facility (IVRRF), 42–53, 42*fig*, 45*fig*, 46*fig*, 48*fig*, 50*fig*, 52*fig*; JCAE, *Radiation Protection Criteria and Standards: Their Basis and Use*, 94–96; as low as reasonably achievable (ALARA) principle, 96–98; *Radiological Worker Training* handbook, DOE, 97, 98–99; Reference Person (Standard Man, Reference Man), 34–40, 46–47, 49, 51, 93, 147n42, 149–50n61, 149n57; relationship between injury and progress, 71–72; responses to radiation exposure risk, 83–86; RESRAD software and assessment of radiation exposure, 28–40, 32*fig*; risk as concept in U.S. public health regulations, 15, 24–26, 98–99, 104–6, 137n20, 143n70, 151–52n73, 171n65; statistical lives in, 51, 59–60; workers' embodied knowledge of exposure, 99–101. *See also* cost-benefit analysis; reasonable harm

robotics, waste management and, 17–18, 20–21, 26–28
Rockefeller Foundation, 70–71, 83–86
Russell, William, 73, 76

Safe Drinking Water Act (1974), 24
Safety standards: ALARA (as low as reasonably achievable) principle, 96–98; EEOICPA's compensation process, 101–6; In Vivo Radiobioassay and Research Facility (IVRRF), 42–53, 42*fig*, 45*fig*, 46*fig*, 48*fig*, 50*fig*, 52*fig*; JCAE, *Radiation Protection Criteria and Standards: Their Basis and Use*, 94–96; Reference Person (Standard Man, Reference Man), 34–40, 46–47, 49, 51, 93, 147n42, 149–50n61, 149n57; regulatory frameworks for acceptable risk, 24–26; remediation and monitoring on Hanford Reach, 107–9, 109*fig*; RESRAD software and assessment of radiation exposure, 28–40, 32*fig*; risk as concept in U.S. public health regulations, 15, 24–26, 98–99, 104–6, 137n20, 143n70, 151–52n73, 171n66; waste reclassification, 127–30. *See also* reasonable harm
Samatar, Sofia, 26, 39–40, 178n12
Saturday Review, 68
Scarry, Elaine, 106
Schieber, Caroline, 97
Sellafield, 28
Shoup, Bob, 117–18
Snicket, Lemony, 52
soil contamination: acceptable toxicity for, 6; at Hanford Nuclear Reservation, 1–2, 3, 138n25; RESRAD software and assessment of radiation exposure, 28–40, 32*fig*
Sonneborn, Tracy, 76
Standard Man (Reference Man, Reference Person), 34–40, 46–47, 49, 51, 93, 147n42, 149–50n61, 149n57
statistical lives (statistical person), 51, 59–60; Reference Person (Standard Man, Reference Man), 34–40, 46–47, 49, 51, 93, 147n42, 149–50n61, 149n57; remediation goals and, 109, 109*fig*; Stuart R. Gunn, Case 0102, 61–66, 64*fig*
strontium-90, 19, 30, 79, 96, 164n59
Sturtevant, Alfred, 73
Superfund Act (CERCLA), 111–15
Superfund National Priority List, 15–16; comprehensive land use plan (CLUP), DOE, 109–15; limits of remediation, 38

Survival Under Atomic Attack (FCDA, 1951), 81, 82–83

Target You (FCDA film), 81
Taylor, Lauriston, 84–85
Tender (Samatar), 26
Them! (film), 117, 177n48
Thezee, Christian, 97
Thompson, Dorothy, 83
tissue simulation, 44–46. *See also* phantoms
torso phantom, 46–47, 46*fig*, 48*fig*
toxic encephalopathy, 93
toxicity: environmental regulations and, 6; Hanford's Long-Term Stewardship (LTS) program, 113–14; as a concept, 8–9, 16; Reference Person (Standard Man, Reference Man), 34–40, 46–47, 49, 51, 93, 147n42, 149–50n61, 149n57; RESRAD software and assessment of radiation exposure, 28–40, 32*fig*
Toxic Substances Control Act (1976), 25
Tri-City Herald, 89, 117
Trump, Donald, 54
tumbleweeds, 115, 116*fig*

Udall, Mark, 90
uranium: Hanford Nuclear Reservation, 88, 135n1; U.S. Transuranium and Uranium Registries (USTUR), 53–57; In Vivo Radiobioassay and Research Facility (IVRRF), 43
U.S. Department of Energy (DOE), 11; aging waste facilities and budget constraints, 19–20, 20*fig*; CERCLA (Comprehensive Environmental Response, Compensation, and Liability Act), 111–15; Clinton administration, transparency and accountability under, 87–90; CLUP (comprehensive land use plan), 109–15; cost-benefit analysis in policy decisions, 25–26; exposure events, response to, 27, 147n40; Hanford Site Cleanup Completion Framework, 113–15; Intercalibration Committee for Low-Photon Energy Measurements, 46; Laboratory Accreditation Program (DOELAP), 44, 49; post–Cold War analysis of waste and contamination, 12; Radiological and Environmental Sciences Laboratory (RESL), 44; *Radiological Worker Training* handbook, 97, 98–99; remediation goals of, 109; response to insect vectors at Hanford, 117–18; RESRAD software and assessment of radiation exposure, 28–40, 32*fig*; waste reclassification, 127–30
U.S. Environmental Protection Agency (EPA), 137n15, 140n31; CERCLA (Comprehensive Environmental Response, Compensation, and Liability Act), 111–15; environmental monitoring technology, 24–25; exposure assessment, 33, 38; regulatory frameworks for acceptable risk, 25–26, 143n70, 147n40
U.S. National Committee on Radiation Protection (NCRP), 14, 74, 84–85
U.S. Naval Radiological Defense Laboratory, 95
U.S. Office of Civil Defense (OCD), 10
U.S. Transuranium and Uranium Registries (USTUR), 54–58, 60–61
U.S. Transuranium Registry (USTR), 53
USTUR. *See* U.S. Transuranium and Uranium Registries (USTUR)

Valkyrie, 17, 26–27
Van Middlesworth, Lester, 69
Viscusi, Kip, 59

Wanapum, 126, 150n64
Washington Post, 89–90
Waste Management Symposia (WMS) 2018, 17–18, 20–22, 26–28
water contamination. *See* groundwater contamination
Watkins, James, 11–12
Weaver, Warren, 73, 74, 75, 76
Welliver, Nancy, 121–26
whistleblowers, 91–94, 91*fig*, 170n52
Wiedis, David, 97–98
women: female phantoms, lack of, 49; phantom testing and breast tissue, 47–49; Reference Person, risk assessments and, 35–40
workers' compensation claims: challenges of, 23. *See also* Energy Employees Occupational Illness Compensation Program Act (2000) (EEOICPA)
Wright, Sewall, 73, 75, 76–77
Wright, Wakefield (Walkie), 100

Yakama Nation, 38–39, 113, 136n9, 150n63, 150n64, 151n69, 151–52n73, 174n31, 179n21; CERCLA and, 113; comprehensive land use plan (CLUP), DOE, 110–15; treaty rights of, 38–39, 110–11, 150n64, 151n68, 174n31, 179n21

Founded in 1893,
UNIVERSITY OF CALIFORNIA PRESS
publishes bold, progressive books and journals
on topics in the arts, humanities, social sciences,
and natural sciences—with a focus on social
justice issues—that inspire thought and action
among readers worldwide.

The UC PRESS FOUNDATION
raises funds to uphold the press's vital role
as an independent, nonprofit publisher, and
receives philanthropic support from a wide
range of individuals and institutions—and from
committed readers like you. To learn more, visit
ucpress.edu/supportus.

Printed in the USA
CPSIA information can be obtained
at www.ICGtesting.com
JSHW081351080124
54999JS00002B/196